The Practical Shepherd
Trials, Errors and Successes
While Maintaining Profitability
by
Abram Bowerman

A division of Mississippi Valley Publishing, Corp.
Ridgeland, Mississippi

© 2024 Stockman Grass Farmer

ISBN: 978-0-9860147-8-9

Library of Congress Control Number: 2023950516

Photos by Curt Holmquist.
Cover and illustrations designed by Steve Erickson, Madison, Mississippi.
Manufactured in the United States of America.

This book is dedicated to the thought, encouragement
and stimulation of shepherds near and far.

Table of Contents

Chapter 1
How I Became a Shepherd

I first took an interest in sheep when I was six years old. It happened this way. My parents were leaving on business for two days and left me at the neighbors'. These were 21st century homesteaders, the man was a dentist and his expendable income was lavished on a large assortment of exotic livestock.

Dave's main interest on his farm was a flock of Gulf Coast Native ewes. Dave's family milked several of these ewes daily touting the benefits of sheep milk. Staying with them for two days I had the privilege of hand-milking one of the ewes and the opportunity to drink raw sheep milk with my breakfast.

Dave had a boy my age. After Chuck and I completed Chuck's morning chores we were free to do as we pleased. We spent those two beautiful summer days playing Indian in the pasture and woods that the sheep called home.

Impressions on young minds are made rapidly and can be long lasting. My two day stay at the neighbors' ended way too soon, but I had already made up mind. I was going to be a shepherd.

My Dad didn't mind spareing some pasture and hay for a few sheep, but I would have to buy the sheep myself. To raise the funds I began to pick up soda cans in local road ditches. Every several months we would take my bags of soda cans to the local scrapyard and sell them. Then I would count my money. It seemed I would never be able to buy a sheep.

One day a new opportunity occurred. A produce auction was started nearby and my family sold our dairy cow herd and began raising tomatoes by the tons to sell at this auction. This gave me an idea. I would grow cherry tomatoes. So after hours (we all helped pick and pack tomatoes) I planted and later picked and sold cherry tomatoes through the produce auction.

In the fall of the year my siblings and I would spend our afternoons picking up black walnuts in our woods. We got $8.00/cwt for the walnuts after they were hulled. The proceeds would then be divided among us partners.

The happy day finally came when I could buy a sheep with my own money. Not to be outdone my older brother decided he was going to buy a ewe also. This turned out to be fortuitous because as I soon discovered, a lone sheep is a stressed sheep, and a stressed sheep is a sick sheep, and a sick sheep is a dead sheep. My brother's investment spared me the pain of learning that lesson with my very first ewe.

Naturally I went back to neighbor Dave and bought a Gulf Coast Native ewe. She was bred to lamb in March. I was nine years old and now I was a shepherd!

That was over 20 years ago. Today I have several hundred sheep. None of them descended from that first ewe, but today's flock was built with the profits and lessons made and learned in the sheep business school of hard knocks.

For several years I took that first Gulf Coast Native ewe and her daughters back to Dave's each fall to get bred. The rest of the year my growing flock grazed the same pasture as our family's milk cow. This arrangement was quite obviously mutually beneficial. The cow ate the grass. The sheep ate ragweed, chicory, cocklebur, Queen Anne's Lace, dandelions. Nearly any weed that grew in that pasture was on the menu at some point of the year. I knew all that because I would spend hours in the pasture watching the lambs eat and play.

The only hands-on help the sheep needed from me was shearing. Each spring I would shear my sheep with a pair of my Mom's sewing scissors! Once the flock grew to a respectable eight or ten head I managed to find a sheep shearer who was willing to come shear my little flock for me.

Kansas rock salt was the only supplement my sheep received for years. Grazing free weeds and salt as the only purchased supplement, every year was profitable.

When I completed the 8th Grade and entered the school of hard knocks, I was determined to kick this whole sheep thing up a couple of notches. First off I bought a big black-faced ram. Next I turned him in with those Gulf Coast ewes August first in an effort to produce January lambs. The results were far reaching.

1. I had to assist delivery of the lambs.

2. Most of the lambs were too stupid to get on the teat on their own so I had to help them learn to suck.

3. With several months of winter left the ewes needed significant quantities of alfalfa hay to sustain milk production.

4. To grow the lambs heavy enough for the Easter market they had to be creep fed high protein feed.

5. Ewe lambs kept as replacements needed their hooves trimmed regularly.

6. My first ram died from internal parasites.

7. Deworming all improved stock commenced.

8. I bought another ram with "improved" genetics. He needed to be dewormed regularly also.

9. Profits shrank. Labor increased dramatically.

10. After three years of this pointless rat race I sold all the sheep.

Reading second hand issues of *The Stockman Grass Farmer*, thinking back over my eight years of lamb production, and considering my desire to eventually go full time in production agriculture without growing wholesale tomatoes, I decided sheep was still my niche, but the business needed to be modeled similar to my first several years with the Gulf Coast Native.

However, one change was in order. This time to eliminate the bother of shearing, I would raise hair sheep. As a side note, to this day I still experience warm fuzzy emotions when I see a flock of landrace woollies. It makes no difference whether they are from the Gulf Coast or the ancient hills of England and Scotland. There is something about landrace breeds that have been shaped by their native environment that is intriguing to me. Landrace woollies are no exception.

Extensive research convinced me that Katahdin hair sheep would work nicely in my environment and with my management style. After obtaining a breeders' directory from KHSI and making numerous inquiries of availability, management practices and prices, I ended up buying a small commercial flock of Katahdin ewes bred to lamb in May.

With a six month break from sheep it was good to be back in business. While it was a little hard adjusting to calling these funny looking hair ewes "sheep" it was quite a pleasure to watch them shedding their own coats in the spring.

Twelve years later most of my sheep genetics can be traced back to those 15 Katahdin ewes that I bought when I was 17 years old.

This book is written from the lessons learned over the past twelve years in the production school of hard knocks.

Chapter 2
Overcoming Challenges —
Reaching My Goal of 400 Ewes

Like many youngsters, I had a vision of becoming fully employed on my own farm, making a living from the land, raising a family, and building my livestock business to a scale that would allow the next generation to get involved.

Extraordinary results are determined by how narrow we can make our focus. I wanted to focus on sheep, not cows, not row crops, not a mechanical forage and feces handling system, just sheep, raised on pasture and thriving in that setting.

With youthful enthusiasm I arrived at the conclusion that 400 ewes would be just about the right size of operation for a young man. But stating my goal only brought protest and ridicule from peers and superiors alike.

How are you going to manage the workload? Where are you going to get the pasture to feed 400 ewes and all their lambs?

Those two questions sank deep into my being. Forage availability and dependable labor are perhaps the most limiting factors for most livestock operations. Of course, if you have deep pockets, labor and forage can be bought. But if the cost of forage and labor inputs equal the value of our lamb crop we would be better off without any sheep at all.

As I began to realize the connotations of labor and forage requirements to support a traditional 400-head sheep outfit and the necessity of showing a profit every year, I remembered the wisdom of my parents.

While growing up if my siblings or I were heard to say, "I can't," my parents would respond with "Where there's a will there's a way." With that phrase ringing in my ears I set out to find the way to accomplish my will.

The labor challenge was easiest to overcome. It didn't take a genius to notice the sheep raised in and adapted to my environment and forage my landscape produced were capable of sustaining themselves and reproducing without my assistance or interference.

I adopted the policy that if a ewe "asked" for help with simple biological functions like hoof maintenance and resisting parasites when the rest of the flock could wing it on their own, that ewe simply got a one-way ride to the sale barn. Since I hired her to work for me I figured we had the right to fire her if she needed me to work for her.

Too often shepherds find themselves spending 80% of their time coddling 20% of their flock. Just get rid of the inferior 20% and if you have the forage multiply the remaining 80% by four. That's about how many ewes you should be able to handle.

By treating the ewes as employees, my workload basically consists of controlling mating seasons, adjusting stocking rates, and managing pasture to optimize carrying capacity.

Forage availability proved to be a bigger challenge. Initially, I bought 40 acres and set out to pay for it by growing wholesale produce in the summer and cutting hedge fence posts in the winter. But 40 acres won't support anywhere near 400 ewes. Land prices were rising so fast that buying more land wasn't going to be an option with my level of income.

Renting pasture was the only economical model that remained, or so I thought. The trouble was I had never known a farm to be available for rent within five miles of home. So I resigned myself to growing vegetables and raising as many sheep as possible on my 40 acres. Then one day while picking green beans, I had an idea that really stemmed back to my boyhood.

When I was 15, I convinced my Dad to let me rotationally graze his beef cows. Within two years our cow carrying capacity nearly doubled. By incorporating some of his hay fields into the pasture rotation the grazing season was extended three months farther into winter. My sheep flock had outgrown the family milk cow pasture and I wanted to continue expansion. So I arranged a deal with my Dad that allowed my sheep to rotate through his beef cow pastures grazing weeds that his cows wouldn't eat. My Dad benefitted from the weed control and my sheep benefitted from the weeds.

Based on that experience we started looking for cattle farmers who were short on labor and long on weeds. A typical offer would be made as follows:

"Let me manage your pastures for you. With rotational grazing management we can increase production and reduce your labor. In trade let me put sheep on your land. 1.25 ewes per cow or basically a 20% increase in stocking rate."

Naturally you need to be diplomatic about making your offer. Don't point out everything they have been doing wrong. You want to sell them your idea, but don't over promise. This is where prior experience comes in handy.

One farm where we did this had a 40 year history of continuous grazing, no break from animal impact summer or winter. The farmer spent his summers making hay and his winters feeding hay. For lack of pasture the calves had to be fed grain for 90 days just to get in shape for the sale barn.

With our sheep/cow/managed grazing scenario hay making and feeding was reduced. Bush hogging became unnecessary thanks to the sheeps' love of weeds. The calves did so much better with plenty of grass available that the cattle owner decided not to feed any grain post weaning. Eliminating that feed bill saved the farmer $20,000. Two years later grain prices doubled and calf prices did not.

There happened to be a television show leasing the property to film wildlife and hunting activities. Managed grazing resulted in better forage and more cover. The result was more wildlife in the pastures. Of course, that really excited the guys from the television show.

Gaining access to forage in this manner offered almost limitless opportunity. Sadly, the country is full of aging farmers who are low on energy and long on resources, albeit the landscape's potential is often discounted by an aging mind. One of the cattle farmers I worked with told me that he was on the verge of renting his pasture out to row crop farmers until I came along with a fresh young vision and lots of energy.

Can it be difficult to get along with someone 50 years older than you? Absolutely, but try arguing with a banker when you're having a bad year and can't make a farm payment! I would rather collaborate with a stuck-in-his-ways, old-school farmer any day. On the other hand, older partners bring valuable insight to the table on market cycles, local history and other things.

This newly discovered forage opportunity created a new problem. I needed more sheep! The first year in this new game we ran a flock on shares, kept every ewe lamb worth her salt and by the time I turned 25, we owned over 400 ewes. Thus, my wife Christina and I were able to quit growing wholesale produce and live our dream of full time stockmanship.

With time land has become available for traditional cash rent and we have also recently purchased another 100 acres, but it was collaboration with cattle farmers that brought my dreams to fruition. Yes, older farmers who were long on weeds and short on labor.

As for my peers who said I couldn't do it, they are still shaking their heads in disbelief. How can one man take care of 400 ewes? My reply — with the help of the sheep I employ! Our flock numbers have been on both sides of 400 head depending on forage availability, but honestly labor is not a big challenge. I average two to three hours of labor per day, depending on the time of year. And work isn't work if you are passionate about your job.

The economic opportunities on these cattle farms was to convert weeds into lamb. Weeds germinate in the spring and grow in the summer. Grass goes dormant in the fall and survives on root reserves until spring. The conclusion was obvious. To make a profit out of this forage opportunity lamb production must synchronize with forage production.

Becoming a full time shepherd is quite easy. Fifty ewes

can keep you pretty busy trimming feet, deworming, lambing in jugs and chasing rogues out of your neighbor's flower beds. The good news is, there is a practical solution for every challenge.

Developing a profitable operation can happen at small scale or large scale. The concept of sheep living on and being sustained by pasture is flock size neutral.

I have found that whether we are working two ewes or a 400 ewe flock, performance always responds favorably as forage diversity increases and when water is kept clean.

Do you want to become a Practical Shepherd?

* List what you need to make it happen.

* Keep an active to-do list.

* When making decisions ask yourself, "Does this move me closer to my goal?"

* Remember, where there's a will, there's a way.

Chapter 3
Small Ruminants Have Big Advantages

Small ruminants such as sheep and goats have some really big advantages, especially on small acreages, under crowded conditions, and in conjunction with urban/suburban micro farms. If you are fortunate enough to have lots of space in the country, don't feel left out. Small ruminants have some really big advantages on the open range too.

Sheep have been domesticated for more than 6000 years, possibly even before cows, horses or dogs. Dogs have often been called man's best friend, but I think that credit may be misplaced. After all, it's the sheep that have provided meat and milk for man since the beginning of time. And before the age of synthetics, who produced the fiber with which man might stay warm? The sheep! After all it was the sheep, not the cow that won the Wild West.

Now that we are all on the same page let's discuss some advantages of small ruminants.

1. High output. As a rule, the smaller the animal, the higher its reproductive capacity. Here is an extreme example. Elephants produce one offspring approximately every five years. Rabbits may produce two to three litters in a summer with six to ten babies in a litter.

Sheep can produce fully twice as much as cows. Birthing twins is normal, weaning 100% her body weight, and those two lambs finish on forage by seven months of age, are sheep traits that are double the production capacity of cows.

2. Early maturity. If properly nourished, ewe lambs may breed at seven months to lamb at one year. This gives her a year's head start over a heifer calf. Early maturity helps mitigate issues with cash flow.

3. Spread out risk. Assuming six to seven ewes can be sustained on the same amount of forage as would be required by one cow, risk of infertility, maternal dysfunction or death loss is now spread out six to seven ways.

4. Broad foraging behavior. Small ruminants thrive on a wide range of plant life. Given the opportunity, they may browse up to 50 plants species per day. Many nuisance weeds and brushy plants will be consumed before grass thereby reducing competition and shifting plant communities toward a beef friendly sward.

5. Easy on fragile soil. With small hooves and being lightweight, sheep don't tear wet pastures up like cows. For that reason sheep can be especially beneficial when grazing newly seeded pastures. Maintaining small ruminants as a part of the general stocking rate will reduce pasture pugging and help synchronize grazers with their preferred forage species.

6. Love rugged terrain. Steep hills are no deterrent to small ruminants and with their small hooves and light weight they won't cause erosion on well managed landscapes. Big rocks, logs or stumps can provide hours of entertainment for the flock, and anyone who has time to watch. Unfortunately in the case of goats any untended vehicles will be turned into an outdoor gym. Sheep are less athletic than goats and the chances of catching several lambs using your windshield as a sliding board is pretty slim.

7. Child friendly. Small ruminants, males excluded, are not likely to severely injure anyone. I got my share of bumps and bruises trying to drive a wether at my friend's house when I was six or seven, and again playing rodeo when I was nine or ten, but never any serious injuries. My Mom would tell me "Go pick on someone your size" when she would catch me teasing one of my younger siblings. For children, that's what small ruminants are, just their size.

8. Easy to handle. Squeeze chutes, head gates, electric prods are all quite unnecessary. Sheep can be manipulated by hand. This is a huge cost saver for small operations. The only equipment really needed to handle sheep is a corral system that allows the operator to effectively pen the sheep or goats tight enough that they are easy to catch.

9. Easy to transport. This one is for micro flocks. Having a few sheep or goats doesn't necessitate a stock trailer for transport as may be necessary with cattle. Most sheep and goats are small enough to fit in a full-sized pet carrier.

10. Small ruminants always milk A2 A2. Individuals with dairy intolerance are finding out they can consume A2 A2 cows' milk. Folks with limited forage resources should consider keeping a couple of sheep or goats in their backyard to produce milk for the family. Unlike cows, all sheep and goats are A2 A2 eliminating the need for DNA testing.

11. Richer milk. Sheep and goat milk compared to cows milk has a higher fat content, higher MCT oil content, is naturally homogenized and in general is more nutrient dense.

12. Always tender. I have never experienced a tough bite of lamb or kid. We have consumed a lot of tender mutton (older lamb). If you ever have lunch with us you will probably be served mutton.

We routinely harvest healthy cull ewes that are two to seven years old. Ewes in good flesh with some fat cover offer a great eating experience. Skinny, stressed, run-down ewes don't have any place on the menu.

When dealing with older animals, slow chilling and appropriate aging the meat are important considerations. Hair sheep are generally supposed to have less of a mutton flavor than wool sheep due to less lanolin in their coat.

13. Easy to harvest. Being small, butchering a sheep or goat is not a big undertaking. I can easily kill and dress a sheep in 30 minutes. Give me a meat saw and I can break the carcass down enough to bag the meat and put it in the refrigerator to await further processing within another 20 minutes.

14. Less storage needed. It is a whole lot easier to sell a whole processed lamb than a whole or half a beef. One reason for this difference is storage. You can put an entire processed lamb in the freezer of a typical refrigerator. Whole and half lambs are more manageable for today's households than are beef whole, halves or quarters.

I'm sure I missed some advantages. Make your own list.

Large ruminants have some big advantages too. As a rule cattle are non-selective grazers preferring grass to forbs, eating everything in front of them and pooping on everything behind them. Cattle will stay productive on low quality grass that small ruminants will only survive on. With their big hooves and superior weight large ruminants break up the soil surface stimulating germination of the latent seed bank.

Properly managed ruminants large and small provide nutritional building blocks for humans, regenerate dormant landscapes and among other things help to mitigate climate change.

Let's keep domestic ruminants out on the land.

Chapter 4
Do You Have a Sheep-Friendly Environment?

Potential shepherds are attracted to this noble enterprise for a variety of reasons.

For some, it's weed control/forage management. For others it's economics. Still others cite emotional equity or aesthetic values. Most shepherds who stay in the business for ten or more years will probably select all of the above.

No doubt ancient shepherds were primarily interested in food and fiber to subsist from day to day. Many a child has been kept alive with milk and slaughtered animals when circumstances sent families drifting. Without refrigeration, having a living reproducing food source that could sustain itself from the road ditch and walk along with the migrating humans was surely a great source of comfort.

Sheep have adapted to a wide range of environments and will do well in your pasture...if your forage resources and production management strategies agree with their constitution.

Let's discuss forage resources first.

Unlike cows, sheep can give birth at one year of age and may produce two lambs every year thereafter. Weaning 100% her body weight should be the norm for a mature ewe. This is twice the production capability of a cow. Of course this outstanding level of production must be sustained by a corresponding level of nutrition.

A sheep can satisfy her nutritional needs during lactation by grazing early succession plants such as forbs and annual grasses. Forbs can be 33% to 45% more nutrient dense than grass. Let me quantify that for you. Every pound of forbs contains 33% to 45% more minerals, vitamins and other life sustaining elements. Annual grasses are highly digestible; rapid digestion allows increased forage intake. These early succession plants drive performance in forage-based flocks.

Late succession, perennial grasses make for great cattle forage and are acceptable feed for dry ewes and yearlings who only need to maintain condition. Sheep will thrive on perennial grasses when they are short and tender, but don't try to keep perennials eaten down all summer. Soil degradation will follow.

Now that we are all agreed that sheep need weeds to thrive, how do we grow them? Consulting for new shepherds who are having production and health issues in their flocks frequently leads to the following statement: "My sheep did really well for the first two years but now the lambs are not growing well and are fighting a lot of health problems."

What has changed? In most cases the sheep have simply eaten up all of their preferred forages and without sufficient soil disturbance new weeds are not germinating. Or if they do, they are quickly smothered by rank perennial grasses. Forbs and annual grasses are early succession plants destined to rapidly cover the soil. The species of forbs that germinate are a reflection of the prevailing soil life, nutrient availability, etc.

As weeds grow up they die and release their specific nutrients back onto the soil surface. Changes in soil life begins to favor grass. Perennial grasses will again dominate the pasture sward. Unless, of course, if the forces of nature or the actions of man set plant succession back again. The healthiest ecosystems contain a mixture of perennial and annual species. And the more species found in a plant community the more "critters" will be found above and below the soil surface.

Nature sets plant succession back in the following ways:

Fire. On a natural under-grazed landscape the fuel load builds up and increases until it finally gets ignited creating fresh highly palatable vegetation after the rains come. In the Old West, bison and other animals would flock to these rejuvenated areas, grazing up to and following the burn line as clearly as if they were contained by electric fence. Or so they say.

Freezing/thawing. Freezing suspends plant growth, fluffs the soil up and may heave plants out of the ground. Early succession alpine meadows are an integral part of many migratory sheep operations. Imagine the growth potential of lambs that are continually moving up the mountains towards the best forage as spring advances upwards.

Floods. Flooding will alter a plant community, but flood prone areas often specialize in cattle forage.

Drought. Drought tolerant species may proliferate during dry spells. Water loving plants will die.

Heavy animal impact. Picture a million bison stampeding across the prairie. I can almost feel the ground shake. Regardless of the species, large herbivores and especially large herds will break up a dormant landscape and stir up the latent seed bank. Your pastures don't have to appear badly pugged to ensure weed germination. Whether we are talking about two cows or 100 cows they are going to disturb the soil surface.

Man can't control freezing/thawing and drought. Lighting a fire may even be illegal during dry spells. And depending on your location and resources, keeping a few cows may not be an option. If such is the case, then plant succession can be set back in the following ways:

Tillage. Shepherds used to recommend lightly disking one-third of the sheep pasture each year to rejuvenate the ecosystem. Of course, you would want to disk a different third each year. Any practice that disturbs or exposes the soil will likely bring on a flush of forbs.

Mechanical clipping. Occasionally removing uneaten, untrampled grass will give new seedlings a chance to grow. Alfalfa, which is really a domesticated, glorified weed, thrives on routine mowing or grazing all the way to the ground every time it begins to bloom, approximately at 30 day intervals.

Herbicide. I am not making any recommendations here, however, chemical burn-down will reset plant succession and allow the plant community to rebuild from the ground up.

Healthy soil should be the goal of every operation, however the more healthy the soil is, the faster plant succession will advance. It is unlikely you will find a natural landscape suspended in an early state of succession located in a temperate environment. Grazing only sheep in a temperate environment is unnatural.

From the Midwest to the African Serengeti, large herbivores have been a part of natural landscapes since the beginning of time. So if adding some large grazers is not an option be prepared to intervene as necessary with tillage or some other tool to encourage continued production of sheep friendly forages.

Now for production management strategies:

As the production manager, it is up to you which season your workforce (the ewes) birth and raise their lambs. The ideal lambing season will vary from farm to ranch and from region to region. In most cases it would be best to lamb in sync with a flush of high quality pasture growth. Here in Missouri we grow very little forage November through March. Trying to raise lambs during this dormant season is cost prohibitive.

Based on the Standard Animal Unit Table, a ewe more than doubles dry matter intake when going from mid gestation to mid lactation if supporting two lambs. Knowing the growth curve of your forage and synchronizing peak forage consumption with peak forage growth will dramatically reduce the quantity of stockpiled and stored forages needed to carry the flock through the dormant season, provided the lamb crop is marketed at the end of the growing season.

I have seen several all-natural flocks busy lambing in mid-winter on low-quality standing grass and supplemented with equally low quality grass hay. Performance was terrible. Many of the ewes didn't have milk. Death loss was extremely high. Employee/ewe turnover was embarrassing. Folks, that is not sustainable. When asked why they didn't control the breeding season, they said removing the rams was unnatural. When asked why they didn't feed some good legume hay to those hard working ewes, I was told it was too expensive. Natural or otherwise, if you can't afford to feed high quality hay in the winter, then control the breeding/birthing seasons to correlate with a flush of fresh green forage. I have never known a ewe to be without milk if she has been grazing fresh spring grass for two to four weeks.

Remember, green forbs and annual grasses are the forages sheep need to produce 100% of their body weight. They eat perennial grasses that are in a vegetative stage and will also browse a lot of woody plants, but forbs are what really drive production.

This basically leaves us with two options. One, grow out lambs when the forbs are growing. Two, mechanically harvest and store the forbs to maintain their quality and quantity for the dormant season, enter monoculture alfalfa fields. My sheep are opting for the first model, growing lambs in the growing season.

Before we end this brief discussion on production management we need to discuss protection from the elements. Sheep have two major defenses against cool temperatures and moisture: a heavy winter coat with lanolin to keep water from penetrating and the ability to build up back fat. When a sheep lambs she begins to lose her back fat and with it her insulation against the cold. When a lamb is born it comes out with a summer coat and very little lanolin to repel water. Consequently winter lambing flocks need a lot more shelter than they would at a milder time of year.

Does one size fit all flocks? No. Every operation has its own opportunities and challenges. Forage options will vary. You may find that your highest quality forage is growing during the winter (eg. winter annuals in a Southern state). If that is the case you may want to lamb in sync with that forage. Just be prepared to offer adequate shelter. And whatever you do, enjoy it. Production is not sustainable without passion.

Chapter 5
Sheep Fences —
Keeping Them Effective

"Good fences make good neighbors." The wisdom in this rural refrain is easy to understand.

No one appreciates the neighbor's livestock trespassing in their lawn or vegetable garden. Nothing strains relationships like a dairy bull slipping in with registered Angus cows or ram lambs from neighbor Jones getting in with a flock of ewes four months before the planned breeding season. Indeed, fences need to be effective.

Fences are divided into two basic categories, physical and psychological. It is important to understand that when sheep see other sheep on the other side of a barrier, their psychological instinct is to physically cross that barrier and join the other sheep. A physical fence is a barrier sheep cannot get through, over or under. Most woven wire fences eventually fail to hold sheep if another flock is on the other side.

If you don't want to risk mix-ups, the best advice is don't place sheep where they can see other sheep. Woven wire perimeter fences are nice but barb wire may work too. I have had good success with existing barb wire fences on farms we rent by placing a single strand 12.5 gauge high tensil wire, offset from the barb wire fence one foot from the fence and 12 to 16 inches off the ground. This scenario is pretty much a physical barrier provided the offset wire stays hot.

Effective psychological barriers all have one thing in common, the sheep believe they cannot cross the barrier. This belief is driven by fear and the barrier remains effective as long as the sheep remain afraid. Water is a good example of a natural psychological barrier. Sheep fear water and if the pool or stream is wide and deep enough to require swimming, it will usually contain sheep. However, given sufficient incentive, sheep can overcome their fears of water.

Case in point would be the drought of 2012, I watched in utter amazement as my flock of ewes and lambs walked into ponds up to their chests in water to graze water plants that happened to be the only green living material they had seen in four months. Indeed, sheep can be very adaptable when they want to be. I also heard of a renegade ram who would swim across a pond to get to a flock of ewes. Those examples illustrate how a psychological barrier becomes ineffective when desire exceeds fear.

When electric fence fails to hold a flock, the cause is usually extreme desire (the flock got hungry or wanted to join another flock) or lack of fear (the fence wasn't hot enough or was too loose, resulting in poor contact). Sheep are not born fence pushers, they get trained to push fence compliments of poor management.

When training sheep to respect electric fences, we usually start by placing the flock inside a physical enclosure. Next we set up our standard polywire fence inside the enclosure the same way we would build it on pasture. But for this training course, the fence is set parallel to the physical barrier and approximately 2.5 feet away. Sheep are curious creatures. It usually doesn't take long for them to check out that funny looking polywire. The result? SNAP you have just trained your sheep to fear polywire, congratulations!

The same training course works for flocks who have learned to plow through your polywire, but it will take longer, maybe a week, maybe two weeks. The flock needs time to settle down and forget that mad rush for the fence.

If your fence is hot (5-6 kv) and you're sure it is properly constructed but the flock gets out two days in a row, you should suspect a rogue sheep who has learned to navigate your fence. Once the rogue is out, the rest of the flock will soon follow in a blinding rush, leveling everything in their path.

The first year I did Management-intensive Grazing with my sheep, a rogue ewe soon taught the rest of the flock to blow through the fence. Giving her the benefit of the doubt, I retrained the flock inside a physical barrier then continued with my regular grazing program. The same ewe taught the flock to get out again. This story repeated itself several times until I realized that rogue ewe would have to find another home.

Today I have a lot less patience. If another rogue ever develops it will end up in the freezer without a trial. Lamb and tender mutton is after all, my family's favorite protein.

After eliminating that first rogue and two more who came with a flock we bought a few years later, I have never had a flock break out again in more than 11 years. Lambs born in the flock are self trained to avoid electric fences within days of being born. Once in awhile a sheep will accidentally stumble over a polywire fence in tall grass. In that case, I'll pardon her and put her back in the flock on probation. How extensive an electric fence needs to be to be effective depends ultimately on how much grazing impact you want to apply.

For permanent interior subdivisions we use three strands of 12.5 gauge high tensil wire spaced from the ground at 12", 22" and 32". On my 40 acre home farm, deer are an endangered species so we use steel T-posts spaced 40 feet apart with pinlock insulators. This allows me to raise or lower the fence to create a "gate" anywhere I want.

Some of the farms we lease have numerous deer. To reduce the risk of the fence shorting out, these farms are fenced with non-conductive posts. Three wires may sound like over-kill, but wire is cheap and long lasting. The bottom wire is left cold 98% of the time. The middle and top wires are both kept hot, the top wire at 32" is primarily for controlling cattle. Hedge or Black Locust posts planted four feet deep make great end posts to stretch the three wires; no brace is needed if the posts are planted four feet deep and are well tamped.

For portable, temporary fencing, I started with three strands of polywire spaced at 7", 15" and 24" and a polypost every eight steps or approximately 22 feet apart. When the sheep began to break out daily I finally upgraded to electric netting. It contained them, but it was too inflexible for my system and much too expensive. Identifying and eliminating that rogue ewe allowed me to switch back to polywire fencing.

Today, we use two strands of polywire spaced from the ground at 7" and 17" supported by blue O'Brien polyposts every 11 to 12 steps or approximately 30 feet apart. Sheep are usually more inclined to duck under a fence than to jump over. If they are required to graze your pasture down short, having the bottom wire at eye level will prevent accidental breakouts. Any polywire used should have no less than nine conductive filaments. A combination of steel and copper conducters seems to work best.

With the help of a geared reel, I can build and then take back up 660 feet of two strand polywire fence in 16 minutes. Best of all, I can easily carry all the posts in one arm and the reel in the other hand. How's that for portable fence?

Some folks are getting along well with one strand of wire but from what I've seen, this requires more polyposts and reduces how much grazing pressure can be applied. In other words, sheep get hungry quicker with less fence to navigate between here and the next salad bar.

Keeping the fence hot day and night, keeps the flock respectful of the fence. I dislike intelligent fence chargers that slow down when they sense a draw on the fence. The fence should be tested daily and if the charge drops below 4 kilovolts when the grass is wet with dew, either unhook half the farm or get the weedeater out.

When selecting a fence charger we always get a model several times larger than the factory rating says my fencing grid should need. Real life conditions are much more taxing than the controlled conditions these units are tested in.

A common misconception is that more joules will produce more kilovolts (or a bigger ZAP). But that is not the case. More joules just means the eletric punch will travel more miles from the charger. As of this writing, cyclops and stayfix both make pretty dependable, durable fence chargers.

Take away points:
* Train sheep to recognize and respect electric fences.
* Don't let them get too hungry.
* Distance flocks to avoid mixups.
* Keep the fence hot.
* Adapt fencing to your needs.
* Keep temporary fences flexible and very portable.

Chapter 6
Water Development Aids Grazing Management

A sheep's water needs are minor if compared to cows or horses. In fact, non-lactating ewes on green forage may go weeks without drinking. Grazing dormant grass buried under a few inches of snow will generally satisfy a sheep's thirst even if the forage is brown.

On the flip side, lactating ewes and growing lambs will drink prodigiously in hot weather. All classes of sheep will increase water consumption when eating dry hay or dead, brown stockpiled forage. Whether water consumption is high or low, the water needs to be clean. More on that in the next chapter.

An improvement in grazing control, managed grazing is usually followed by a need for better water distribution. One low cost, inflexible method used is a central water source with all paddocks watering from the hub or center of the wheel.

The biggest problem with this central water method is the fact that in order for the livestock to graze the far end of the paddock out by the rim of the wheel they have to walk back across the pasture they grazed the last six days to get a drink. With rapid growing conditions pasture regrowth may be sufficient in three days time to stimulate regrazing.

This regrazing thing isn't the worst thing in the world, but when this scenario is repeated season after season, year after year, the front half of the pasture will deteriorate from over grazing and the back half may deteriorate from a lack of impact and selective grazing.

Decentralized water allows for increased forage production and utilization on the whole farm. When I first embraced the concept of managed grazing I chose to haul water for the sheep for two years. This was done with a 275 gallon tote on a hay wagon. I would pull the wagon into a new paddock, hook the sheeps' water tub/float system to the tote and the sheep would use the wagon as an additional shade source in conjunction with their shade mobiles.

After two years of hauling water I bought a 2400 gallon water tank at a farm sale and set it on the highest hill in my sheep pasture. From that vantage point water could be piped to every corner of the farm via one inch polyethylene pipe. The pipe was laid under electric fences as much as possible to keep livestock from walking on it and guard dogs from chewing it up.

This system was cheap and provided ultimate water distribution, but the whole system had to be drained before winter to prevent freezing. As a result, winter water availability limited my grazing options unless it snowed, and the storage tank had to be refilled at regular intervals.

One year I teamed up with a neighboring cattle farmer. He had weeds and I had grass. We both needed what the other didn't want. My sheep would spend the summer on his weedy cattle farm and his herd of cows would graze my grassy sheep pasture, a win-win relationship indeed.

After seeing how fast 100 cows and 90 head of yearling calves drained my 2400 gallon storage tank I realized my water system would have to be upgraded to keep our grass/weed swap viable.

Always looking for ways to harness nature and reduce energy consumption, I figured the ultimate water system would be a pond positioned in a place that would allow water to gravity flow to every pasture. The problem was out here in the boondocks ponds are always built at the bottom of a valley. This allows for a bigger pond but greatly limits where that water can flow in a pipe.

Regretfully I made the same grave mistake on the first two ponds we built. The good news is the ideal elevated locations for ponds were still available. Obviously I like ponds.

Today when we build a pond we select a location as high on the landscape as water can be stored economically. From that vantage point water can flow downhill and to some degree along the contour.

We generally use 1.25 inch polyethylene pipe buried 30 inches deep to deliver water from the pond to the pasture. My friend, Dennis, has been doing this for 30 years and says he has never known a pipe to freeze. Another advantage of buried water lines is its convenient protection from equipment and livestock.

To access the water we install a riser/plasson quick coupler valve at regular potential watering sites. The riser is contained in a six inch pvc casing with a cap at ground level. This riser system costs a fraction of a hydrant and does not require a drain field. Plus, sheep and cows can't damage these risers the way they would with a typical hydrant.

When I want to hook a portable water tank to the 1.25 inch delivery system, I take the cap off the pvc pipe, stick a male plasson coupler attached to a garden hose into the plasson valve located 16 inches below the surface. Presto, the stock water tank is attached to a huge storage tank called a pond. Best of all, no pump is needed. The pond fills naturally when we get a heavy rain and we are storing the runoff water for later use.

To keep our portable water tanks open in the winter we crack a small valve inside the tank to allow a steady flow of water to circulate through the tank. By trial and error we determined that 40 gallons/water/hour flowing through a 20 foot garden hose into a small poly tank will keep the whole system open. Severe cold and wind may require a little more water flow, perhaps 70-80 gallons/hour. The length of the garden hose and size of the tank ultimately depends on how much water is needed to keep everything thawed.

The extra water flowing into the tank 24/7 simply runs downhill, creating a glacier if it is cold or a soggy pasture if temperatures are mild. Keeping those conditions in mind when planning watering points and fence layout is a worthwhile exercise.

The simplest way I have found to adjust the flow rate of the seep in the tank is to plumb a small quarter turn valve into the system ahead of the float valve. The quarter turn valve can be adjusted with the weather to sustain the desirable flow rate.

Just turn the valve off in the summer, unless you want to irrigate your pasture.

Developing a decentralized water system creates untold flexibility resulting in better pasture utilization all around.

Chapter 7
Water — Keep It Clean

One year I rented a farm similar to the one described in the previous chapter. The only obvious difference being water pressurized by an electric pump and bladder system. Water distribution was fantastic being pressurized rather than gravity flow. Water was literally available everywhere include hilltops. I remember thinking, "Great, at least we have good water."

The diversity of forages on this farm was awesome and the cattle that had been on it previously for the most part seemed to thrive. Yep, everything was in place for a productive summer. As expected, the sheep loved the forages. During April and May the flock gained weight, lambed and seemed to feel right at home.

June came. The rains quit. Temperatures soared. The forage dried out and the sheep began to drink a lot more water than they had in April and May. Next the flock started scouring. We were puzzled. Then it cooled down. The scours ceased.

That cycle repeated itself over and over until I finally realized the periodic heat waves were somehow causing the scours. But, my other two flocks were getting along just fine in spite of the heat. So, I dug deeper looking for the culprit. The problem was hiding in the water tank. Actually the culprit wasn't hiding. Its presence was quite obvious. The wonderful pressurized water on that farm was filthy dirty. The sheep hated it but there was nothing else to drink.

The cows had survived that smelly dirty water, but it was killing my sheep, slowly but surely. Sheep, as I learned, can't handle a toxin load that cows will survive. Diarrhea or scours was the sheep's way of expelling that huge overload of toxins.

So what's wrong with pressurized water? The answer is nothing. The problem was where the water was being pumped from, which happened to be the bottom of a pond 14 feet below the surface.

As temperatures rise in the summer, the surface of the pond heats up. Oxygen levels in the surface water decline and the pond turns over. This stirring action turns the dirty water at the bottom into filthy stuff that smells a little like a lagoon.

Where in nature do you see animals sticking a 10 foot straw down in a pool of water to slurp up the dregs? In nature animals always drink from the surface. Sunlight kills harmful bacteria in water and among other things sunlight purifies and structures water to a depth of possibly three feet.

Back to those poor sheep. When I realized my mistake we revamped the water system to draw water near the surface, but we paid for our actions for 12 months.

A friend once told me about a similar ordeal that happened to his own flock. Summer came, the sheep started scouring. He assumed the problem was internal parasites. Operating on that belief he dewormed the whole flock and moved them to the other side of the farm to avoid re-exposure to a parasite-infected pasture.

Cattle were grazed on the first pasture to clean up the parasite load. A year later the sheep returned to the first pasture. Before long they were scouring again. After some consideration the shepherd remembered the well water supplying the paddocks his sheep were scouring in, was said to be high in salt. The obvious conclusion was to give the flock another water source. He did that and the scours dried up.

The morals of these two stories is sheep need clean water and flock-wide scours should always leave the water supply suspect.

I have always wondered why so many of the ponds in our area have a plugged water line. One day I got my answer. We were building a new pond. The dozer operator had dished out the basin and asked for a hand to unroll the new water line through the middle of the dam. Next he drove a post into the the pond floor and wired the pipe to the post. Then he got his Dewalt drill and proceeded to drill 1/4 inch holes in the pipe.

After a bit I said, "Hey, how low are you going with those holes?"

"Within a foot of the bottom," he said.

"Why?"

"Because that's how NRCS says to do it."

For the record, NRCS wasn't sharing the cost on this pond, and I wasn't about to end up with a plugged waterline when the pond began to silt in. We argued for a minute until I convinced him that he wasn't liable to NRCS or anyone else for the consequences of drawing water closer to the surface.

That incident occurred before we realized how important clean water is. Today when we build a new pond or revamp an old one the pipe is extended slightly above the surface. Next we attach a weight to one side of the pipe and a float to the other side. With a little adjustment this suspends the pipe 12 to 18 inches below the surface, which keeps it from catching floating debris from the surface. And yes, no water is drawn below a depth of three feet.

Sheep fear getting their feet wet. I have never known them to loaf in ponds like cows will. This natural good behavior allows us to water the flock directly from ponds and streams without ill effects. Some folks say sheep fear their own reflection in the water. True or otherwise, sheep can be fenced with water deep enough to swim. This natural tendency to stay dry protects riparian areas.

Water plants such as sedges, cattails and arrowleaf help to hold pond banks in place and serve as a filter to keep the water cleaner. If your pond is home to frogs, fish or pollywogs, it is probably clean enough for livestock to drink, but be sure to draw water near the surface because that is only natural.

Water delivery systems and water quality are among the best investments we have made in our grazing business. The sheep express their appreciation daily via greater performance.

Chapter 8
Sheep Need Shade

The fact that sheep sometimes need shelter from the sun was impressed on me when I was 17. Up until then my sheep were pretty much set-stocked on various permanent pastures, always with some kind of natural or manmade shelter. I just sort of took this shelter for granted.

Shortly after my 17th birthday I bought my first flock of Katahdins only budgeting enough money to buy some portable electric fence and a one year supply of Kansas rock salt.

Witnessing the improvment in pasture quality and production brought about in my Dad's beef operation by simply implementing managed grazing two years earlier I was determined to do the same with my new flock of hair sheep.

Alas, with very limited fencing resources and regular paddock shifts always including some kind of permanent shade shelter in each paddock was impossible.

I failed to take shade into consideration in my original production plan and operating budget. The sheep were obviously suffering in their enclosure. Just as I thought I would have to give up managed grazing in favor of shade Bowerman creativity came to my rescue.

I found an old Canfor tarp salvaged from a bunk of 2 X 6 lumber and transformed it into a tent with four open sides and five feet of clearance to encourage air flow. The sheep loved it! The tent was supported by a wooden truss and two legs at each end with a 3/8 inch rebar attached to the legs to stick in the ground. Baler twine served as guy wires to brace each leg. Baler twine was also stretched between the two trusses to help support the tarp. The tent could be set up in 10 minutes, was easy to transport and was made from a salvaged tarp, baler twine and some scrap lumber. Fortunately my flock couldn't care less about the appearance of their tent. It sheltered the flock that whole summer, though it did get a little crowded as the lambs grew.

Even though the tent worked quite well I determined to build shade on wheels for my second year. After all, the flock was growing and I didn't want to move two tents every day.

A wagon running gear purchased from a construction company that was going out of business provided the base for my first shade mobile. The gear was built ultra heavy with a wide wheel base, features necessary for supporting scaffolding on construction sites. The extra weight and width of the gear is beneficial to keeping the shade mobile from flipping over in the wind.

That first mobile shade unit is still intact and occasionally in use in our operation today. As flock size grew I continued to build more mobile shade units as needed. Some of those structures have been reduced to twisted scrap metal, a result of being too tall, too wide and the gear too light.

While mobile shade gives you some control over manure distribution it has the major disadvantage of depreciation. Given the option, all else being equal, livestock will choose tree shade over shade mobiles. Tree shade has a greater cooling effect than lifeless shade structures.

It's generally accepted that all classes of livestock perform better when there is the shelter of trees and shrubs in a landscape. It's also well understood that livestock performance increases when their diet is not limited to non-woody plants or one plant family.

When I realized the depth and significance of that last paragraph we began to channel some of our energy toward developing and incorporating natural shade resources. Efforts have ranged from planting trees to thinning existing timber stands, creating silvopasture. The advantages are many.

Shade. Maintaining thermal equilibrium improves

animal performance and well being. A comfortable animal will spend more time grazing. Shade can make the difference between comfort and misery.

Protection from sunlight. Sometimes sheep just need to get out of the sun even if they aren't hot. Lack of pigment in the skin and hair will send pink-skinned sheep under shelter sooner than dark-skinned sheep.

Various weeds can make sheep photo sensitive, sometimes resulting in a condition called "big head." The head and ears swell up to gross proportions. Shade access is necessary to alleviate the condition. In severe cases the flock may need to be penned in total shade.

Sheep penned on a snowy field on a bright sunny day sometimes go snow blind from the glare. I have often seen a flock seek shade for relief from the light on bright, sunny, snowy days.

Mineral pumps. Trees pull minerals from deep within the soil releasing them on the surface in the form of decaying leaves. Sheep will eat significant quantities of mineral-rich tender young leaves, buds and bark from trees and bushes.

Shelter from wind and rain. It's hard to match the calm that is felt on a stormy day or night when you approach a timber belt. A significant amount of rain can be caught by tree tops and guided to the ground by the twigs, branches and trunk. Evergreens are par-excellence when it comes to shelter.

Trees create a micro climate. The level of diversity often found where pasture meets woods is hard to match. Many plants, birds, and animals, sheep included, thrive in this micro climate.

Natural pest control. Many trees and shrubs offer natural parasite control when consumed internally and serve as an insect repellant when animals rub on them or lounge under trees.

Improved microbiome. Diversity in the diet is positive. Sheep enjoy acorns, buckbrush berries, apples, etc. Increased food options improves a sheep's and all livestock's ability to fill her mineral and vitamin needs.

When planting trees I like to start with early succession, fast growing species. Once they are well established late succession or climax species can be introduced. Sheep will eat and rub on little trees, seemingly trying to destroy the very element they so much enjoy. Protecting small trees is a necessary measure for a few years.

The fastest way to develop a silvopasture is to convert an existing timber stand. My goal is to maintain 30% to 50% shade cover. We usually start by harvesting any mature trees. Next I go through and select the trees we want to keep.

Variety is resilience so we keep as many species present as possible. Obviously preference is given to individual tree health and potential commercial value. A maturity range from three inches diameter to five years pre-harvest may also be desirable.

The first time I saw a landowner do a project like this nearly ruined my enthusiasm for converting existing woods. This landowner started with a climax forest of hickory and white oak. The only trees that remained were tall and spindly, approximate six to ten inches in diameter. Shade cover was reduced to 20% with 80% open in a single thinning with most of the shelter suddenly removed. These tall spindly trees were whipped around in the wind and many of them broke off. After ten years the surviving trees finally started to grow and bush out. See how learning from others' mistakes can shorten your own learning curve?

The winter of 2022-2023 in between moving sheep and writing this book I opened up seven acres of timber to create silvopasture. As the remaining trees grow and branch out a future harvest will be needed to maintain enough sunlight at ground level to grow healthy forages. With that future harvest in mind a lot of healthy larger trees were left to protect the younger ones while they adjust.

To reduce biomass I cut and sold all the firewood from the tree tops and from the younger trees that needed thinning. The remaining brush we piled and burn as time permits. The stumps of species sheep generally won't browse were treated to prevent regrowth. Elm, mulberry, birch and other species that the sheep like were left untreated to produce succulent forage for the sheep. That summer the sheep began to graze the area periodically.

It has been my observation that silvopastures require periodic thinning to maintain the microclimate effect and prevent the area from growing back into a climax forest.

While it is true that deep forests generate better quality lumber, semi-open transition zones between woods and pasture produce much more fruit and nuts, nature's way of trying to propagate trees. This transition zone is where life gets exciting.

Chapter 9
Buying Sheep — Principles to Consider

Whether buying your first sheep or expanding your current flock, here are some basic principles I try to follow that you may find beneficial to your endeavor.

* Phenotype is more important than genotype. Phenotype is easily seen. It is the basic structural and muscular composition of the sheep. Phenotype, or you could say body-type, determines to a large degree the animal's performance or lack thereof. Genotype is the genetic make up of the sheep.

In pasture based systems, phenotype is much more important than genotype. The breed you choose doesn't matter. The phenotype of the breed does. Short, deep, close-coupled animals gain weight easily. This is necessary for production on forage.

A good ewe will have a big belly, wide butt and will stand taller at the hip than the shoulder. This gives her the appearance of walking downhill. Ewes should appear refined, feminine.

A good ram will have a big chest and stand taller at the shoulder than the hip giving him the appearance of walking uphill. Rams should appear masculine and rugged.

* Maternal versus paternal breeds. Maternal breeds are prolific, a 200% drop rate should be the goal. Milk production should be adequate for raising two healthy lambs. Maternal instincts are strong, if selected for. Selection in true maternal breeds targets traits that support reproduction and longevity. Maternal breeds are usually better adapted to forage-based production than are paternal breeds.

Paternal breeds are breeds that have been selected for maximum growth, muscling and carcass traits. Rams from these breeds can be used as terminal sires in a crossbreeding program with maternal breed ewes to maximize profit per animal on slaughter lambs. All offspring from a terminal sire should be harvested for meat.

It is possible to run a profitable operation without the use of terminal sires. Using a terminal sire locks you into butchering all the lambs, eliminating the option of retaining or selling breeding stock.

Perhaps the best use for a terminal ram is to cover the ewes that you don't want to keep replacements from, thereby identifying their offspring and increasing the lambs' market value.

* **Naturalized genetics.** All else being equal, the closer to your home you can buy your sheep the better. If you can't find a suitable flock in your region then look for sheep in an environment that is similar to your own. It is important that the sheep selected are adaptable to your area.

* **Rational decisions.** Wannabe shepherds who wake up in the morning and decide to go out to buy a flock of sheep or rapidly expand their existing flock because the grass is getting away from them often end up suffering from buyer's regret syndrome. Finding a few sheep for sale on Craig's List or at the local sale barn is not a rational place or way for new shepherds to get started. Until you know how to sort the bargains from the wrecks buy private treaty from well established flocks.

Whenever possible buy directly from a producer who has been in business for a few years. Visit several operations before you make a decision and if necessary get on a waiting list. If you have to wait a year, well, you will have 12 months to get fully prepared.

If buying mature ewes, getting them two to three weeks after they wean their lambs might be a good idea. In this post weaning scenario bad udders will be obvious. Most ewes who failed to raise a lamb will be super fat and any ewe who is producing too much milk will be quite thin. Buying ewes at weaning time allows you to see them in their work clothes.

* **Similar management.** How do you plan to manage your sheep? How much individual handling does your labor budget allow? The answers to these questions will serve as guidelines when searching for your future flock.

Buying sheep from a hands off flock does not guarantee you will be able to stay 100% hands off, but it is a good start in that direction.

* **Flock size matters.** The bigger the flock, the greater the pressure will be on maternal instincts. All else being equal small flocks always outperform large flocks. Said another way, performance falls as sheep numbers rise. Increased flock size equals increased stress, competition, disease exposure and daily exercise. The bottom line: big flocks exhibit lower performance and higher mortality rate but the survivors are more productive if moved to a smaller flock, often out performing the sheep that are native to the smaller flock.

I picked my first Katahdins from a flock of 70 head and for several years we thought they had pretty good maternal instincts, but when my flock size grew from 50 to 100 to 150 head I suddenly found that we needed to cull a lot of ewes that lacked sufficient maternal instinct to function in a large flock. Flock size matters!

* **Ewes versus ewe lambs.** Mature ewes have proven themselves. They offer greater cash flow for the short term and being fully developed you can see what you're buying.

Ewe lambs potentially have more years ahead of them and if sourced from a low input flock should develop into low maintenance ewes.

Availability and price are probably the most important considerations. If I am buying sheep and the seller is offering his ewe lambs and six-year-olds for the same price I would choose the lambs every time. That being said, a two- to three-year-old ewe may be worth half again the value of a ewe lamb.

* **Expect adjustments.** There will always be a cost when you move animals. Some producers suggest a 25% death loss/cull rate within 12 months of relocating a flock. I think

that is unnecessarily extreme. I am going to suggest that if you are moving healthy young sheep into your area from a similar environment with similar management and you do your best to ease the adjustment the death loss/cull rate should be less than 5%. The problem is we always get excited about trying out these new genetics and push them too hard before they settle into their new home.

Give your new guests time to feel at home. Treat them to your best forages and they will work for you for years to come.

Chapter 10
Feeding Your Sheep —
When to Make a Paddock Shift

Being a benevolent shepherd, it is my pleasure to feed a flock of hungry sheep. Throughout most of the year the best way to feed a sheep is out in the pasture. To maintain pasture production and sheep performance, the flock should be given a fresh paddock at frequent intervals.

Since the only consistent thing is change, when and how often to make a paddock shift is a decision that should be made adaptively based on the context of your sheep, pasture, labor, season and prevailing weather.

Let's take a brief look at each of these five elements:

1. Sheep. The flock needs fresh pasture once they have depleted their current supply. Do you want to keep your sheep on the highest plane of nutrition possible? Watch what they eat first when they enter a new paddock. These plants are the trigger species. Take note of the top three to five trigger species and when none of those plants are left in the paddock it's time to move the sheep.

The trigger species will change with the sheep's preferences throughout the growing season. Continued observation will be necessary. "Trigger species" is the same concept as "trigger height," the commonality being that when the trigger species are all consumed or the grazing residual reaches the trigger height a paddock shift is triggered.

If I turn the flock into a paddock expected to last 48 hours but find they are hungry in just 24 hours, they need to be moved. On the other hand if 48 hours of grazing has not eliminated the trigger species and the sward has not been reduced to a trigger height there is nothing wrong with leaving the flock for another 24 hours.

The sheep just need to be well fed, that's all. The left paunch, the area between the hip and last rib is a great monitor for gut fill. A full gut and a lack of interest in changing paddocks is an indicator that the flock is not hungry. If 90% of the flock is up, browsing the dregs of the pasture and they answer your call to move they may need a fresh paddock.

2. Pasture. The pasture sward almost always benefits from shortened graze periods, daily shifts are better than weekly. Shorter graze periods result in better manure distribution and more uniform animal impact.

Each paddock will vary in stand density and species composition, so paddock size will need to vary to accommodate your desired stay on that paddock. The first several species your sheep target when they enter a new paddock is an indication that they would like more of those plants in their diet.

If no trigger species are present — sheep show equal preference for most of the sward — then it will be necessary to determine a trigger height.

A paddock shift is generally due if the grass has been grazed down to the second leaf or 50% of the plant.

3. Labor. Both physical and mental effort are a required part of livestock ecology. Labor per animal unit varies between flocks and forage programs. Large scale flocks almost invariably can reduce labor costs by intensifying grazing management to extend the grazing season. Small scale flocks may reduce labor by extending paddock size and then feeding hay during the dormant season. Of course if you enjoy what you are doing labor really isn't work.

At the end of the day labors' part in livestock ecology and ultimately determining when to make a paddock shift is availability.

Offering fresh pasture daily may be desirable but if I'm not going to be available for three days we just give the sheep a three-day paddock and the sheep stay happy so long as the paddock is oversized enough to allow access to fresh un-trampled forage every day.

Sometimes in preparation for a busy week I'll set up several extra paddocks. Then when the sheep need to be moved all I need to do is let them into the next paddock and move the salt feeder and water tank.

4. Season. No one is more seasonal than stockmen, but too often seasonality is embraced by default resulting in an increased workload for man and beast. In other words too often we work against the seasons, but I digress.

Seasons have many implications for grazers and graziers alike. Let's look at a few of the important factors affecting potential paddock shifts.

* During seasons of peak forage growth significant regrowth can happen within three days of a grazing event. Ideally the sheep would be moved every three days or oftener to prevent them from grazing any regrowth.

* During lambing season paddock shifts should take place midday because most births happen between 5 am and 9 am and 5 pm and 9 pm.

* Lambing season efforts should be geared toward animal comfort. Graziers who are excessively compulsive about frequent paddock shifts would do well to back off a little during lambing season. Twice or thrice daily moves for a lambing flock is an invasive practice. Once a day or even every second day is frequent enough for young lambs.

* During summer heat I prefer a late afternoon/evening paddock shift to allow the flock to get oriented before the next day's heat.

* During winter a late afternoon shift is often ideal because the frost may be melted off the forage. Also evening grazing will result in peak rumination in the middle of the night creating rumen heat during the wee hours while the flock is bedded down.

5. Prevailing weather. An expected change in weather often sends me out to move a flock sooner than planned. Forage is best consumed dry. Moving before a thunderstorm rather than after is desirable.

In the winter if the flock is trapped on a windy knoll with plenty to eat but a blizzard comes shrieking in I'm going to move the flock to a more sheltered area or at least give them access to some shelter. I will sleep better knowing my sheep can get away from the wind if they want to.

The same principle applies to the summer sun. If it's been cool and cloudy and the flock doesn't have access to shade but the sun comes out then they need to be moved no matter how much forage is left. Never mind a little wasted grass.

In summary, a paddock shift is due anytime animal performance will drastically suffer from negligence. Forage production and forage use efficiency usually rise with more frequent moves, but must be balanced with labor and sheep comfort.

Now, go see if your sheep need to be moved.

Chapter 11
Designing a Sheep Corral

Nothing adds to the ease and pleasure of handling sheep more than a well designed corral system.

While the size and design of the corral needed will vary with the flock size and management practices, there are three features that should be observed in planning a corral system.

1. Make it large enough to easily accommodate the number of sheep involved.

2. It should be conveniently located in regards to pasture, water supply and accessible in wet weather.

3. It should include at least one large pen for holding the flock, either a chute or a narrow alley for separating or handling individual sheep, and two smaller pens in which to put the separated sheep.

Given the option, I usually set my corrals up under shade trees so that I can work sheep at any time of the day in the summer. The best way to avoid excessive handling and coddling of sheep is to mitigate stress. Stress preceeds almost all disease so taking a little time to make the trip through the corral quick and pleasant will reduce the need of handling facilities.

The basic needs of a corral system is to allow the convenient removal of market animals, male lambs and cull ewes and to control mating seasons, the removal of four-month-old ram lambs or else breeding rams at the end of the breeding season. Beyond these basics the corral may be used for vaccinating, castration, ear tagging, deworming, hoof trimming, etc., if you are into all that.

For flocks needing a lot of TLC a chute is usually desirable and if you have a chute you just as well have a two-way sorting gate on the end of the chute. If you build your own chute, build it with adjustable sloping sides, narrow at the bottom and wider at the top. This makes it harder for sheep to turn around in the chute.

With my long standing policy of staying "hands off" I have come to avoid the use of chutes except for sorting purposes such as breeding season. I may wish to split several hundred animals into two or three groups of specific bloodlines to be mated to certain rams. Also since most of my pasture land is leased, building extensive handling facilities is not a priority.

For rented farms I have come up with a basic corral design that answers the needs of a commercial flock. While I don't consider it portable, it is easily removable if I lose access to that farm. As a side note anything attached to the property becomes the property of the land owner unless protected by a contract. So when you write your grazing contract put a clause in there allowing you to build and to remove a corral.

The design is as follows:

1. A round or oval shape depending on topography and shade availability.

2. An outer fence of four feet tall cattle panels supported by a T-post every eight feet.

3. An inner circle made of hog panels reinforced with one inch tubing. Being in a circle these are freestanding.

4. The inner circle is set five feet from one side of the outer circle and 14 feet from other side of the outer circle creating a lopsided donut.

5. Two hinged hoop gates are set five feet apart on the side where the alley narrows to five feet. This is the "Bud Box."

6. A third hinged gate opens from the Bud Box into the center pen or hole in the donut.

7. An eight foot gate is placed in the alley approximately 25 feet before the Bud Box. Groups of 20 to 30 sheep can be trapped in this section of the alley prior to entering the Bud Box.

8. A 14-foot reinforced hog panel is set at an angle in the alley 20 feet from the other side of the Bud Box. This pen is for the sheep that have been worked.

9. As the sheep cycle through the Bud Box any animals being retained are shuffled into the center pen or hole of the donut. The rest pass on through back into the outer circle.

10. As the work progresses the 14-foot panel mentioned in #8 can be advanced around the outer circle to make room for the sheep that have been worked and to push the unworked sheep around closer to the Bud Box.

11. Usually I also build a catch pen two or three times the size of the corral. This is constructed of cattle panels or woven wire. On rare occasions I use electronet but it is not intended for high pressure so be careful how you handle the sheep.

My preference for working sheep in a Bud Box is based on the ability to see conformation, action and soundness. The sheep are brought forward in groups of six or eight at a time and as you are walking along behind them any potential problem animals can be spotted before they ever get confined in the Bud Box.

Sometimes if a cattle corral is present on a farm I am leasing I just attach woven wire to the alley and narrow the alley down in one place to five feet, bring in a couple of hoop gates and get to work. The critical thing is to be able to tighten the sheep up enough that you don't have to chase them around a pen to catch each animal.

If I have more than 150 head in an alley of any kind I break them into groups of approximately 100 head with additional gates to relieve the crushing effect.

The easiest way to move sheep forward in the alley is to open the gate you want them to pass through and then walk down the alley toward the sheep. They will flee past you into the next pen.

If you go behind the group and try to ram them forward you will scare the ones nearest you and the sheep in front may not move at all.

Helpful hints. When working with sheep keep in mind:

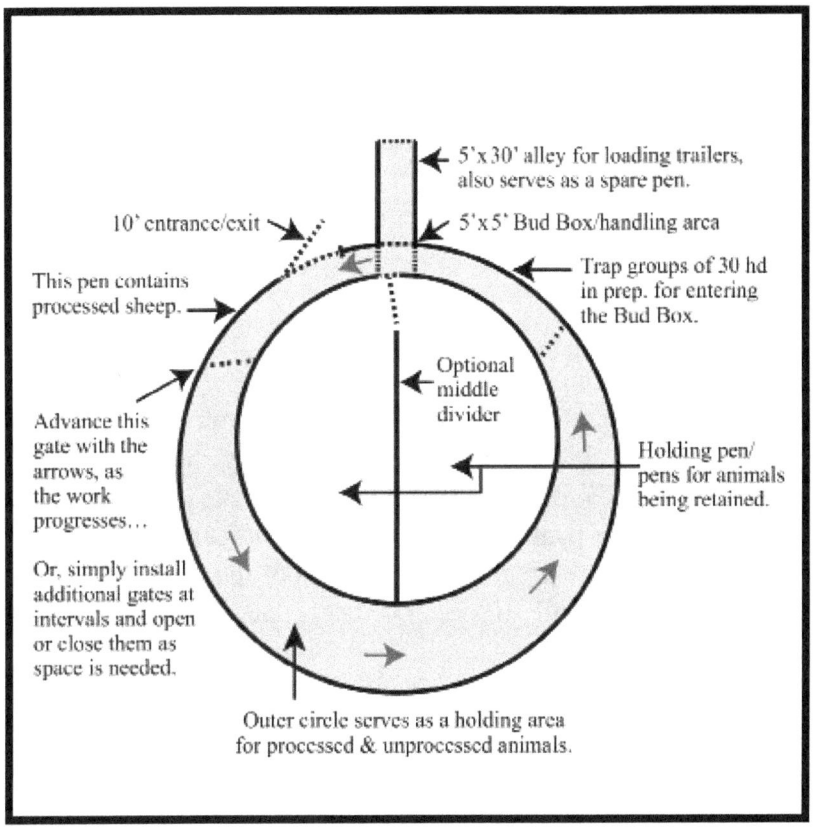

5'x30' alley for loading trailers, also serves as a spare pen.

10' entrance/exit

5'x5' Bud Box/handling area

This pen contains processed sheep.

Trap groups of 30 hd in prep. for entering the Bud Box.

Optional middle divider

Advance this gate with the arrows, as the work progresses...

Or, simply install additional gates at intervals and open or close them as space is needed.

Holding pen/ pens for animals being retained.

Outer circle serves as a holding area for processed & unprocessed animals.

* Sheep prefer to move uphill. Use this to your advantage when penning in the corral.

* Individual sheep stress, so move multiple animals in the Bud Box at the same time.

* Stay calm. Allow plenty of time for the job and use this opportunity for close observation.

* Open all the gates in the outer circle before driving your sheep into the corral, resulting in more space to enter.

* Once the flock is penned, the easiest way to move a group of animals forward is to open the gate and walk toward the sheep. They will flee past you into the next pen.

* The combined area of the loading alley and Bud Box will hold enough sheep to fill a standard width x 24 foot stock trailer.

Constructing this area with solid wood will reduce the risk of animal injury during handling.

Chapter 12
Sheep Psychology —
Working with Sheep

Understanding sheep psychology and applying your knowledge in a skillful manner when handling sheep will reduce stress levels for all involved.

Here are some basic principles of sheep behavior:

* Sheep are gregarious, creatures of the flock. Where one goes they will all go sooner or later. Sheep confined by themselves stress.

* Sheep fear noise, unfamiliar surroundings, unfamiliar objects placed in familiar locations, strange dogs, stranger people and sheep don't like to get their feet wet, even a small creek may be deemed un-navigable.

Sheep are easy to move from confinement to an open area and prefer to move up hill and are difficult to move into dark buildings.

* If the front animals move along, the sheep in the back will follow with little or no pressure.

* Any sheep that is off over the hill by itself is probably either lost, blind or really sick.

Flocking instinct varies between breeds, but as a rule, when you begin to gather a flock they bunch up. Driving a large flock across an open landscape is not a one man job. Unlike cows, sheep can't seem to stay focused on the direction you start them and will suddenly race off in another direction, but before they get there they will change course and take off who knows where.

Sheep love to play follow the leader. Once they figure out who the leader is and where that leader is going, they will

54

follow until something distracts them and another leader takes off in a different direction.

If you simply want to lead the flock into an adjoining paddock, that is easy. They soon figure out the routine and will follow a shepherd on foot, on a horse or on an ATV. Anyone can make a good leader provided they are trusted and can hold the attention of their flock.

For long drives with lots of potential distractions, there are basically two options for controlling the flock.

1. Fences on both sides of the course to keep the sheep following you.

2. The shepherd leads the flock and a dog or else two athletic youngsters keep the flock following the shepherd.

With a little practice, small flocks of 20 to 30 head can easily be herded and penned by one man without a dog. In this scenario the sheep form their own leader and the herder guides the flock from behind and either side. Temporary electric fences can be set up to help guide the flock.

I suggest that mastering herding on foot should precede any action on horseback or with a herding dog. If you can't get a desirable response by yourself when afoot, it is unlikely you will do any better with a horse under you or a herding dog helping. Practice is the key to successful handling and practice starts at ground level, with you.

In my mid-teens I frequently drove my small flock of sheep from one field to another on our 300 acre farm. No fences or dogs were used to move the flock. One year my neighbor and I agreed to ship our lambs together to a sale barn 70 miles from home.

My neighbor had a flock of approximately 30 ewes and lambs. To corral them he enlisted the help of another small scale shepherd. After four hours of fruitless effort they gave up and found five more people to help pen 30 head!

When my neighbor showed up at my place with his lambs and saw my 20 ram lambs in a little makeshift corral by the driveway, he asked, "How did you get your lambs there?"

I responded that I had let them out of their pasture back

over the hill and drove them up there and herded them into the pen. Incredulously my neighbor asked, "How long did that take you?"

My reply, "Ten minutes."

"How much help?" he wondered.

"None. Just a little experience."

That practice herding sheep has never ceased to be a benefit when I am working with sheep or other livestock.

Bud Williams, famous for his stockmanship ability once said, "Moving and working animals should be a quiet, pleasant experience for both animal and handler." Bud was right. It can be pleasant and quiet, but too often incompetent stockmen strain family and neighbor relationships while stressing the livestock they are handling.

Here are three steps you can take to help mitigate stress for all involved, including the sheep.

1. Practice. Learning to gauge flight zones and animal/flock response to your gestures and physical position will improve your ability to corral and handle your sheep. When you have time turn a flock loose in a big field. This is just for fun. Practice gathering and driving the flock, but limit your exertion to walking. If the sheep race off without you, fine, let them go. Keep walking. You will catch up eventually.

When the sheep settle down and accept your general presence, practice driving the flock and then holding them against a ditch, fence or other barrier. Practice should not exceed two hours. You don't want to exhaust the sheep.

Next, put the flock through your corral system, just for fun. This exercise will benefit you and the flock. Once familiar with the corral, they will flow through the system much smoother. If your corral system includes a chute, pen the flock, open all the gates, and allow them to file out through the chute on their own. Sheep have good memories, so a quick, painless trip through the chute will improve compliance next time through.

Our preschool children practice herding ducks into a pen every evening. Their herding skills develop rapidly from this regular practice and I see it manifested in the way they respond to the sheep when I let them help with a sheep drive. Herding ducks is also a great pastime for youngsters.

2. Allow plenty of time. Everything goes better when the schedule isn't cramped. I would rather start my day at 4:00 a.m. and get done working sheep by noon than sleep in and find we are up against dark getting done. However, some days just don't go so well and schedules get cramped; but don't let it affect you. You can only operate so fast. Trying to push beyond optimum will slow you down.

If you know approximately how many sheep are in the flock you're working and tally the head count as they pass through the Bud Box, one hour into the job should leave you with a good idea how long it will take to finish. If I realize there is not enough time to complete the job, the next step is to eliminate one or more processes on the day's agenda.

If none of the day's objectives can be dropped, then I go to Plan B — letting everyone involved know that we are going to be late. With eight hours of advance notice, everyone can adjust their plans accordingly.

3. Forgiving attitude. While I try to stimulate everyone, including the sheep, to their topmost performance, it is still necessary to leave room for error. When you feel like blowing up, stop and ask yourself, will this failure matter in five hours, five months or five years? Can it be addressed later?

The time lost from the mistake is already lost. Taking time to retaliate on a ewe or hold a shout down with your helper will only lose more time.

Designing your corral system to contain all the sheep until the whole job is done can eliminate a lot of stress. If a sheep escapes or accidentally gets sorted into the wrong pen, it can be re-circled through the system.

Corrals designed to release processed animals directly out to pasture will give you a nervous breakdown and the sheep that are still contained in the corral will stress every time they see one of their number make the exit.

One time I was two-thirds done working 300 head in just such a corral when a big ram lamb saw his mother exit the corral, headed for timber. This ram lamb was in the main holding pen, waiting his turn to be processed. Seeing his mom and half a dozen other sheep heading for the woods, he stuck his head through the cattle panel enclosure and started lunging. Before I could get to him, the end of the panel came loose and most of the unprocessed sheep escaped, joining the processed flock on pasture. I was castrating ram lambs that day and only had one third of the flock left to go, but now I had to start all over again, catch every ram lamb, set him up on his butt and see whether he had been clamped.

Lessons learned? Building the corral stout enough to avoid blow-outs and design the system to keep all sheep contained until the job is done.

No matter how good a stockman you are, the layout of your corral will have a big impact on your work efficiency. Number one you want a system that sheep will flow through quite readily. When given the options, I prefer a curved alley over a straight one. Sheep move through a curved alley better, perhaps because they can't see what's in front. Maybe because they think they are returning into the flock, but probably because it allows them to keep one eye on the handler.

Any long alley should be split with gates to divide the flock once it is in the alley. These gates eliminate the crushing effect of a large flock caused by all of them rushing toward one end.

When you want to move sheep up the alley open the gate and, staying on one side of the alley, walk toward the sheep. They will flee past you into the next pen. Trying to drive 100 head from behind will scare the ones nearest you and the ones in front may not move at all. When working sheep, remember, noise louder than conversation is detrimental to your objective.

Perhaps for many shepherds the most stressful job of all is loading sheep on trailers. Many a truck driver has commented on how quickly and easily my lambs load. The conversation goes something like this.

"Wow, that was easy. They jumped right on. I trucked

a load of lambs for Joe the other day and we had to drag them onto the trailer, one at a time."

All the local truckers are convinced that my sheep are unusually intelligent. They may be right, but I believe loading technique deserves most of the credit. This is where understanding a sheep's psychology comes in handy.

What are sheep scared of:
* Strange objects (trailers).
* Strange people (truckers).
* Loud noises. No shouting. Please turn off your truck.
* Fear moving from light areas to dark ones. Open the big trailer door.

Trying to get sheep to walk past a trucker, stick its nose inside a dark trailer that probably smells like cows not sheep, and jump in single file will often result in difficult loading.

Here at Still Waters I ask my truckers to back up to a five-foot alley and stop soon enough that the entire rear gate can be opened, which allows for plenty of light. Next we extend the alley out to the trailer with the help of two portable gates. These two gates take all the stress out of aligning the trailer with the alley. They are simply wired to the back of the trailer and extend from there back along the alley where again they are wired or chained for support.

Once everything is secure the sheep are released to move toward the trailer. I walk back through them and they head for the trailer and jump on five abreast. If they stop at the trailer instead of jumping on, a gate is advanced up the alley behind the flock to hold them near the trailer. Then I move around to the front, pitch one or two lambs into the trailer and before you can snap your fingers the rest of the flock is following them onboard in a blinding rush. The result, they push the first lambs all the way to the front of the trailer. I close the trailer gate, unwire the gates used to extend the alley and we are off to the races.

It's all a matter of technique.

Chapter 13
Learn to Recognize Healthy Sheep

Whether buying sheep, selecting replacements from your own flock or measuring the flocks' response to management decisions, learning to recognize healthy sheep will pay dividends.

Astute shepherds will notice a positive/negative response to dietary changes within two to three days time. Sheep dairies have the distinct advantage of measuring fluctuations in milk yield every 12 hours.

Fortunately milk production is not the only sensitive gauge for flock health. Performance potential is reflected by hair/wool, hooves, activity or lack thereof, manure, etc.

What can a lamb tell you?

When a lamb is born it should exhibit at the least a two-inch-long umbilical cord attached to the belly. Don't pull it off, it will dry up and fall off naturally. Failure to exhibit an umbilical cord at birth heralds problems in the future. Or you could say suggests problems with your forage program in the past.

Healthy young lambs may begin to bounce and play within hours of birth and definitely within two to three days. Within 10 days the newborns will form troops, racing up and down paths. Lambs that are doing well will gang up and play king of the mountain on every rock, log, stump or pile of dirt in their pasture.

Flocks that have been suffering from inadequate dietary carotene will exhibit a higher incidence of difficult births, retained placentas and a high mortality rate, especially for lambs that are less than nine days old. Ewes or lambs with contracted tendons is almost a guarantee that carotene levels in the diet were inadequate during gestation.

Healthy lambs nursing ewes with optimum milk production will exhibit a fat glob or double chin on the throat where the head attaches to the neck. New shepherds often panic the first time they see this fat glob and ask me if it's goiter or bottle jaw, thinking their lambs are wormy. Truth be known, I have never seen a lamb succumb to parasites while it had this double chin. A fat glob under the throat is a good omen.

What can the fleece tell you?

A sheep's coat hair or wool is a pretty good indicator of what is going on inside the body. And ultimately the level of production she may be capable of in the near future.

I like to keep a few black sheep in every flock. They make the flock easy to spot on a snow covered field and among other things add character. One winter I noticed a striking difference in two of my black ewes. One was jet black, the other had a dull faded coat that looked brown.

Both ewes were about the same age, the same size, on the same diet and in similar body condition. Three months after I noticed the faded brown coat both ewes lambed.

The ewe with healthy jet black coat birthed two lambs and raised them both. The ewe with the dull faded coat also had two lambs. One lamb was a premie and died within two hours of birth. The other lamb appeared normal but died from pneumonia 30 days after birth.

Regardless of color, white, brown or black, a sheep's coat should have a bright, healthy appearance. A ewe in full fleece may have a duller coat than she will after shedding or being sheared, but a full fleece should still be bright and alive.

If I was given only one selection tool I would choose the hair/wool coat indicator. Hair on the head and throat are the most rapid indicators of change or improvement of environmental conditions, also indicating a sheep's ability to adapt to your environment.

Shedding dead hair will begin on the head and may occur at any time of the year. Spring and fall are the most normal occasions triggering a shedding event. The spring shed off is the most pronounced.

Hair on the poll of the head is the most visible, sensitive barometer on sheep accurately reflecting the well being of the animal, its hormonal balance and its nutritional status.

Poll hair will vary in length, texture and posture; however, with practice, a glance at the poll will give you a quick idea of how that sheep is feeling. Long hair standing up as if electrified is a strong indicator of internal disturbance. Shorter hair laying close to the skull is a good sign.

A ewe's poll hair is generally fine and silky in texture. The hair on the poll of a healthy fertile ram will be short, bristly and lay close to the skull. Any ram with silky poll hair standing up as if electrified is probably sub-fertile.

The posture of the poll hair on ewes and rams can change dramatically within two to three days in response to stress and rumen pH. Remember I said hair on the poll is a sensitive barometer.

What can hooves tell you?

The outer wall of the hoof should be smooth and shiny. A dull, bumpy ribbed outer wall indicates the sheep's inability to meet her nutritional needs in your environment.

Sheep with hoof walls heavily ribbed are unlikely to pass their full genetic potential along to their offspring.

So what? Given the option, choose the animal with four smooth shiny hooves. If all else is equal, dark hooves are better than white hooves, but healthy white hooves are better than compromised black hooves. Take time to notice hooves. They support the sheep industry!

What can lips and eyelids tell you?

Sheep with a healthy red blood cell count will have lips and a nose with a healthy pink color. Black nosed sheep will still have pink lips on the inside.

A sheep with ashen or gray lips, nose and inner eyelids is not healthy. She may be fighting a parasite load or it could be something else, but be advised that she is not thriving. Moral, select for naturally healthy pink lips.

What can manure tell you?

Manure quality should be monitored every time you go to the pasture. I don't mean collecting a sample to send off for lab analysis, you can do this yourself.

Healthy sheep manure will look sort of like raisins. Depending on moisture and maturity of the forage your sheep have been eating, these raisins may be firm little balls or they may be pressed together to form a bigger glob. In the case of the glob, individual raisins should still be in evidence.

On the other hand, unhealthy manure will look like a lava spill or cow manure. Lava manure from sheep is proof of metabolic disturbances and perhaps amplified by internal parasites.

What can observation tell you?

Research is pointless without keen, unbiased observation. Frankly, livestock make ideal researchers proving by their own performance or lack thereof what works and what doesn't.

Shepherds with a keen sense of observation will develop a benchmark over time for flock health, ear carriage, head carriage, animal behavior, gut fill, manure quality and the profile of the poll hair, all things you will come to instinctively observe and monitor without even realizing it.

What do problems tell you?

Part of learning to recognize healthy sheep is noticing problems and signs of future failure.

Mastitis. Ewes with a mastitis infection spend a lot of time laying down, are slow to move and may appear lame. The udder will be inflamed. If only one side is infected, the udder may appear lopsided. Infected udders may feel hard and hot.

Culling ewes who have had mastitis is top priority, even if only one side is affected.

Anemia. Internal parasites should be suspected if the flock at large becomes anemic. The symptoms are:

1. Ashen/graying color of the lips, nose and inside of the eyelids.

2. A puffy upper lip. This symptom may appear to be an overbite at first glance but closer examination will reveal a swollen upper lip protruding noticeably beyond the bottom lip. A puffy upper lip is an easy way to diagnose anemia in black nosed sheep and can be seen from 20 feet whether the nose is pink or black.

3. Lack of endurance is also a common symptom. In severe cases of anemia caused by parasites a small trek requires much exertion due to the lack of red blood cells to carry oxygen to the tissues of the body.

Lameness. Contracted tendons, foot scald and footrot all have one thing in common, they make exercising painful.

Lameness is typically a result of malnutrition or poor circulation. If most of the flock is sound those that are not are probably not adaptable to your environment and forage resources. Chronic cases should be culled.

If historically the flock has been sound but the majority of the flock pulls up lame, changes in the pasture sward, water quality and winter feed should be considered.

Broken mouths/gummers. Broken mouth is the term applied to a sheep that has lost one or more of its adult teeth. A gummer is a sheep that has lost all of its teeth or has worn her teeth down to the gums.

After a ewe turns six years old her teeth should be checked annually. A ewe with an unsound mouth will struggle to maintain body condition and raise healthy lambs.

Tooth retention varies drastically in different breeds and flocks. Loss of adult teeth may begin as young as five years of age. Other ewes may retain a solid mouth into their teens, thus tooth retention is a trait worth selecting for. I find it highly heritable.

Let's wrap up this discussion.

* Every rapid rise in disease is our fault.

* Sheep have ways they can demonstrate their adaptability.

* Weight gain is good. That's what we get paid for.

* Keep in mind that even though a ewe may maintain her body weight over winter, it doesn't prove she's covering her nutritional needs. Just look at homo sapiens.

* Problems are always preceded by deficiencies.

* If the minority of the flock is struggling, cull them. If the majority is struggling, change your management.

* The ultimate proof that a sheep is happy and adapted is reproduction.

Chapter 14
Culling the Flock

Culling is one of those things that will take care of itself if we are slow to step up to the job. Turned loose in the wild, any sheep that is a little slow on her feet is going to be taken out by predators. Lameness, obesity and weak respiratory systems will all lead to the same result.

Even in a pasture with a coyote-proof fence backed up by reliable guard dogs, predators will continue to cull your flock for you. Quiet thieves go by the names of bacteria, viruses, internal and external parasites, etc. Nature is always busy recycling animals that are not happy or are not fitted to their environment. Ultimately, this culling action frees up space and valuable forage resources for adapted animals.

Eliminating all predation on your flock is impossible. Fortunately, with a little knowledge, experience and observation, it is possible to take the role of predator yourself. Let me state that clearly. You can be the predator in your own flock and for your own benefit.

Culls may not bring a lot of money but $75.00 is seventy-five dollars and will have more value to your sheep business than the composted remains of a dead sheep. Please don't take my numbers to heart. The principle is what I want to convey. You could even make culling an investment budget, 10 culls X $75.00 = $750.00 to invest into a freeze proof water delivery system to extend grazing opportunities in the winter.

A lack of forage suited to a sheep's metabolism will lead to a higher than normal cull rate. Promptly eliminating culls will free up valuable forage resources for healthy productive animals.

Failure to cull the flock will end in death loss. Bad udders will leave you with dead lambs, same thing goes for lack of maternal instinct. And old age will take every ewe sooner or later. It's easy to think "she may give me one more lamb," but a cull ewe in good shape will often sell for almost as much money as a fat lamb. So go ahead, sell the dysfunctional ewe and keep another ewe lamb. It probably will give you a lamb the first year. Plus, a ewe lamb will appreciate in the next 12 months, whereas an older ewe will depreciate. And a ewe lamb has the distinct advantage of consuming less forage than a mature ewe.

I remember one year when sheep prices rose sky high. Many producers held onto ewes they normally would have culled. A year later, prices took a nosedive and sheepherders sold two years of culls in one year. This action drove a depressed market even lower. As a result, cull ewes were sold for less than half of what they would have brought the year before. Moral: what goes up comes down, the higher it flies the farther it will fall.

Weaning time is usually the biggest culling event of the year. Four months of lactation will divide the pros from the cons. However, don't limit your culling to weaning time. The best time to cull is your first opportunity to conveniently catch a dysfunctional animal.

Here is a basic list of symptons to cull for:
* Bad udders/mastitis.
* Chronic lameness.
* Anemia.
* Broken mouth/gummers.
* No bag at weaning (didn't wean a lamb).
* Ill thrift (any symptoms of ewes' health apart from or below the flock average).

Any ewe without a bag at weaning time is a cull candidate, especially if she is obese. Being fat and milkless is a sure sign she never lambed or lost her lambs soon after birth. The problem that killed her lambs could have many different names from acidosis to or anything in between. The point is, she didn't wean any lambs and the rest of the flock did. In my experience, any ewe that fails to wean healthy lambs one year has a greater than flock average chance to fail in future years. I have also found that the ewes who fail to wean healthy lambs have a higher than flock average chance of dying suddenly.

New producers often ask if there is a certain age they should cull ewes. My answer is yes. The age a ewe begins to show symptoms of future failure. Sorry, I know that's not really what you wanted to hear. Culling every ewe when they reach a certain age would eliminate some potentially productive ewes that still have several years of potential left.

Tooth loss is often the only reason an older ewe may need to be culled. Tooth retention varies greatly in sheep. I have seen broken mouths as young as five years and I have seen ewes who still had a sound set of teeth at 13 years. Keeping replacement ewe lambs from ewes that are over eight years old is a great way to improve tooth retention and other characteristics of longevity in your flock.

Now that you're whipped into a frenzy to do a little culling consider the following. Next time you drive your flock down the road take note of the laggards. Those slowpokes are very likely needing a ride to the sale barn. Lameness, mastitis, obesity or a respiratory weakness will quickly leave a sheep behind. A half mile drive is usually sufficient to separate the pros from the cons. For some shepherds cull criterion is based almost solely on the outcome of a fast one or two mile march. At the end of the drive the gate to the pasture is shut before the stragglers arrive and the culling job is complete when the last straggler is loaded onto the trailer.

If driving the flock down the road sounds like too much work, then just watch closely every time you move your flock into a new paddock. Any sheep that is consistently slow to enter a new pasture may be a cull candidate.

With close observation followed by decisive action, you can be the predator in your flock. Now go make a list of potential culls.

Chapter 15
Calendar of the Flock

Designing your production calendar to fit your labor, forage, cashflow, and infrastructure resources and objectives is paramount for successful production. Among other things, my objective as a shepherd is to produce lamb, and to produce it at the lowest cost per unit as is practical. Yes, I believe in least cost production. Just don't confuse Cheapskates with Least Cost Producers. There is a big difference! Cheapskates try to starve a profit out of their business. Least Cost Producers identify and eliminate unnecessary costs from their production model.

Here at Still Waters, the production year begins with the breeding season. The whole production calendar is predicated on this important date.

After taking a quick look over the flock for potential culls, the rams are turned in with the ewes. A ewe's estrus or heat cycle comes every 17 to 19 days. To ensure every ewe has had two chances to breed, rams should stay with the flock a minimum of 38 days. I usually leave our rams with the flock until 60 days post lambing. This saves on chores.

A healthy flock exposed during their natural breeding season (October — December) should have a 100% conception rate on their first estrus cycle. The resulting embryos are just floating in the uterus for 30 days after conception. During this first month rough handling or stress can result in loss of one or more embryos. Granted, if a ewe conceived twins, losing one embryo will result in a single lamb the next spring. If both are lost the ewe will simply go to estrus.

Forage quality should remain fairly high during early gestation to avoid reabsorption of one or more embryos. After the embryos attach to the wall of the uterus (around day 30) the pregnancy becomes fixed and we are on our way to lambing season.

Gestation length for sheep is approximately 150 days or basically five months, but varies. Variable gestation length seems to be linked with breed and phenotype. However, I have seen the average length of gestation for a flock vary up to seven days from year to year. Ewes experiencing malnutrition will lamb earlier to allow for increased forage consumption.

A short effective breeding season is culminated in a short lambing season. 99% of the ewes will usually lamb in a 19 day window. Other than ear tagging, ultimately we don't interfere with the ewes during lambing season. Instead we focus our energy on keeping the flock safe, comfortable and well fed. If a ewe fails to mother-up, she is culled later in the summer. We have stuck to this cull criterion long enough that it is unusual to have a ewe reject her lambs.

When the first lambs born are 60 days old we run the flock through the corral and castrate the ram lambs with emas-culators. This method of castration is bloodless and is lower stress more than banding. Letting the lambs grow for 60 days before castration reduces the risk of urinary-calculi later in the finishing stage. This is also a good time to vaccinate against blackleg, tetanus and redwater. All three can be given in one shot.

Sixty days post lambing is also the time to remove the mature rams if they are still present. Fertility is lower during the summer for both rams and ewes but all the same, some of the ewes will likely breed 75 days post lambing if a sexually mature ram is present. The result would be more lambs at the beginning of our regular breeding season. Lambing every six months is not sustainable in my environment.

Ram lambs can reach puberty by 130 days of age. If any ram lambs have been left intact then they need to be weaned at four months of age to protect the ewes.

The longer lambs are left on the ewe the better they will do. If the ewes are only lambing once a year, there is no good reason to short-wean the lambs unless you need to destock. I never wean our replacement ewe lambs at all, consequently it's not uncommon to see three or four generations of females bedded down together. Wethers are also left on the ewe for her to wean when she gets around to it. The difference in lamb performance between natural weaning and forced weaning is amazing.

Four months post lambing (when you are weaning intact ram lambs) is a good time to cull the ewe flock. If we have plenty of forage and wish to fatten the culls before selling them then we just take note of which sheep need to be culled. Notching the ear tag on cull candidates allows a cull to carry her own record and makes for positive easy sorting in the future.

Well fed lambs on efficient ewes should weigh approximately 50% of their dam's weight at four months. If forage quality remains high, these lambs should reach two thirds of their dam's weight by the time they are five to six months old.

Research has shown the ideal finish weight for a lamb is 64% of the average mature weight of the ewe, from dam and sire breeds. Thus lambs can reach their ideal degree of finish within the growing season and be marketed directly off the forage without serving time in the feedlot. I have found that once a lamb reaches the finish weight, growth nearly stops. Consequently, well finished lambs should be marketed unless a significant improvement in price is anticipated later in the year.

Ewe lambs can be bred at six to seven months of age. By then they should weigh two thirds of their dam's weight. Breeding at seven months old keeps your ewe lambs on track with the rest of the flock. I don't sweat over whether a ewe lamb is ready to breed, if she is, she will, and if she's not, she won't. Let them sort it out. If a ewe lamb is physically developed enough to lamb at one-year-old and does not, there is the very real risk of her developing a fatty udder before she lambs as a two-year-old. Plus, raising one lamb as a yearling gives her some experience and she will likely perform better as a two-year-old.

Nutritional needs will vary across the year, but they are highest 30 days pre-lambing to 30 days post-weaning, a period of six months.

As the production year draws to a close and we are approaching the breeding season, the flock can be "flushed" to improve fertility. To flush the flock simply allow them to be more selective in their grazing patterns. Basically a flush equals a rising plane of nutrition. Of course, if fertility is already high, flushing may result in a bunch of unwanted triplets the next spring.

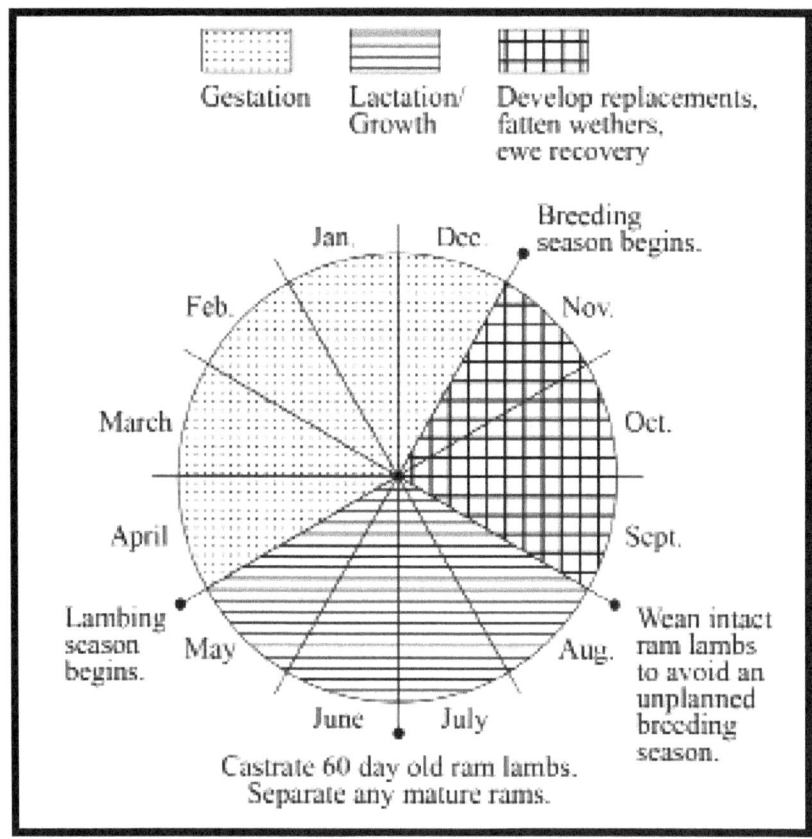

In this example of a production year, nutritional needs will be greatest April 1st (30 days before lambing) to September 30th (30 days post weaning), a period of 6 months. Production should be synchronized with your best forages and other considerations.

The 4 Seasons: Stay Focused

Here in the Midwest we are fortunate to have four distinct seasons. Synchronizing production with our natural forage chain and other considerations has lowered our cost of production and improved our quality of life. Remember Ecclesiastes' injunction: "To everything there is a season, and a time for every purpose under heaven."

Spring is the time for new life to begin. There is nothing like abundant fresh green grass to stimulate milk production.

Summer is the time for growth. Forbs all come along in a flawless chain to provide the nutrition lactating ewes and growing lambs need.

Fall is the time for a sheep to build up back fat. Growth slows as the animal begins to put on fat instead of building muscle. This is simply a natural preparation for winter.

Winter is a time for sheep to survive and gestate. For shepherds it's a time to monitor, plan and meditate.

Chapter 16
Planning a Successful Pasture Lambing Season

Watching 100 ewes drop 200 lambs in 18 days' time gives me a thrill. With adequate planning I can lamb out 400+ ewes by myself without working abnormally long hours. Perhaps I should rewrite that last sentence as: with adequate planning 400+ ewes can lamb themselves, leaving me free to design the grazing activities of the flock.

The key word is "planning." Someone will always plan the lambing season. Planning begins approximately five months and one minute before the lambs are born.

The following is a true story I witnessed of a shepherd who let others do most of his planning. Names have been changed to protect identity.

Tom lost a grazing lease and was forced to sell half of his ewes. Knowing that bred ewes often sell better than open ewes, Tom proceeded to expose the ewes he planned to sell. Being exposed to the rams in mid-September the ewes began to lamb mid-February. In the meantime, Dick bought the bred ewes, expressing his opinion that he sure didn't want to buy open ewes.

Dick was excited when the first few lambs were born on pasture in mid-February, but a series of winter storms soon forced Dick to move the flock into his small barn. Under those crowded conditions, ewes and lambs were constantly getting mixed up until Dick finally set up some lambing jugs for the fresh pairs. Dick was worn to a frazzle by the time he was done lambing his flock of less than 100 ewes!

74

Dick decided to lamb in April the next year. Lambing season rolled around, cold and wet. The grass had hardly begun to green up but the ewes wouldn't eat any hay. On this low energy diet lambs were born weak and the ewes were so busy looking for something to eat that they hardly paid any attention to their lambs. Dick's family soon found themselves busy playing foster parents to a whole bunch of lambs.

Next Dick resolved to bump the lambing date out to May first the following year, but alas, being busy making hay, he was slow getting around to weaning his ram lambs and "they" planned the next lambing season for January.

Sore mouth had established itself on the farm the summer before without any major consequences. Unfortunately these January lambs contracted the sore mouth virus when they were two weeks old. Without the forage diversity that is available in the summer, milk quality was lower reducing the immune response of the lambs. Instead of little bumps on the lips like summer born lambs get, these lambs got huge crusty scabs covering the whole mouth. The results were devastating. To make a long story short, Dick had had enough and sold all his sheep.

There are two basic models of lamb production. Both require planning.

1. Shed lambing in the dormant season. Usually substituting quality pasture with expensive legume hay or other stored forages and concentrates. This model is labor and capital intensive, but allows you the option of easily assisting the whole process of reproduction.

2. Lamb on pasture in sync with a flush of green forage. This requires sheep with strong maternal instincts. Labor is minimized because the sheep harvest their own feed, birth their own lambs and "mother up" on their own.

While there may be reasons to implement the first model, depending on forage availability and access to premium hay, my interest and experience lies with the second model.

Always impatient for lambing season to begin, for several years I kept moving the breeding date a few days earlier in the year, but when I looked at survival and performance data the last lambs born always outdid the first lambs, catching up with and passing them inspite of being 18 days younger. This told me the flock was lambing too early.

A ewe should be on green forage three to four weeks pre-lambing. Good green alfalfa hay works, but pasture is cheaper. In my part of the country, spring green up usually occurs four weeks before the last frost. Incidentally, lambs born after the last frost were always the most vigorous at birth. Plus, these later lambs didn't have to endure as many cold rains or risk a late snow. A young lamb can endure a lot of bad weather if it is well nourished, but this starts with ewes being well nourished, which requires an abundance of high quality forage.

My fears of lambing late were:

1. Heat and flies. Plenty of shade fixed the one problem and maintaining parasite resistance prevented the other.

2. I thought ewes would have **trouble finding their lambs in the taller forage** of late spring. As it turned out, the ewes being on peak nutrition with plenty to eat everywhere they looked didn't move around as much when they lambed and exhibited stronger maternal instincts. The ewes seemed to return instinctively to where their lambs were left.

Bringing all this together, the conclusion was obvious; lamb after last frost! Folks cite markets as the reason they lamb in the bleak months of January and February. I cite costs of legume hay, lambing sheds, and labor as reasons why I wait to lamb until spring.

* While shelter is necessary, the back side of a hill or shelterbelt of trees is usually good enough for dry ewes.

* Legume hay can and should be kept to a minimum during mid-gestation.

* Finally, by planning ahead, lambs can stay safe and warm in the womb until growth quality forage is cheap and abundant.

The ideal lambing date will vary for every flock based on region and resources. Access to cheap alfalfa hay or free annual cover crops could potentially change when I choose to lamb. Creating an enterprise budget for each production model and comparing the profit margin will tell you when to lamb. Whether lambing indoors or outdoors in spring or fall, summer or winter. Planning ahead will help you meet the challenges as they come. Plan the lambing season, or the rams will plan it for you!

Chapter 17
Pasture Lambing Tips

The purpose in pasture lambing is to avoid the cost and inflexibility of lambing sheds, to reduce feed costs, and to reduce labor.

Ewes with strong maternal instincts are a must for pasture lambing to succeed. While maternal instinct certainly has a genetic factor, the nutritional status of the flock also plays a critical role. Starving ewes are thinking about survival, not reproduction.

The quickest, cheapest way I have found to improve maternal instinct and newborn intelligence, is to postpone lambing until three to four weeks after spring green up. In my area, this corresponds with last frost, virtually eliminating the risk of a late snowstorm. A grazing plan that takes lambing season into consideration is time well spent. The objectives of my annual plan run something like this:

* Station the sheep on a farm big enough to sustain the flock until lambing is done. I don't want to drive a lambing flock down the road.

* Plan the paddock rotation in such a way that the sheep can always drift into an adjoining paddock.

* Avoid crossing ditches with steep banks during lambing season. Use polywire to fence out ditches that the sheep can access the rest of the year.

* Build plenty of flexibility into the grazing plan. Don't be afraid to use plan B, C, or D. Implementing a plan is basically taking stock of your resources and leveraging them for success.

Once lambing commences, avoid the temptation to rush

out there and cuddle those lambs. I usually have three flocks of ewes lambing each spring and never step foot inside the lambing pasture of two of these flocks. The third flock I disturb daily by tagging the newborns and removing one lamb from every set of triplets. These "orphans" are easily peddled around the neighborhood. More on ear tagging later.

The best advice is to stay out of the lambing pasture. Any ewe worth her salt can birth her own lambs and mother up on her own. I know how tempting it is to help that presumably lost lamb find its mom, but you risk stirring up other new pairs and ending up with 10 lost lambs!

I absolutely avoid anything that might cause excitement in the flock, including taking strangers out to see the sheep during lambing season. While the sheep may not mind if they stand at the edge of the paddock, my guardian dogs always set up a terrible ruckus when strangers first show up and especially so during lambing season. The dogs will place themselves between the flock and the perceived "threat" barking and running back and forth until the ewes have grouped up and moved to the far end of the paddock.

It's this grouping and moving action that I don't want. Under normal circumstances, a good ewe will stay on her lambing bed for 12 hours for every lamb born. This gives her and her lambs time to bond. The instinct to stay on her lambing bed replaces the need of lambing jugs. Pasture lambing and lambing jugs just do not synchronize.

I have been ear-tagging lambs at birth in my home flock for 14 years. This is a time consuming process that is not justifiable for flocks intended to produce terminal market lambs. Most of the offspring from this flock are either retained for reproduction or else sold as breeding stock. But still, my excuses for tagging look pretty flimsy. I want to know how many lambs are born, give the ewe a maternal instinct rating, potentially know if every lamb born makes it to weaning age and finally I like to be able to track lineage.

Tagging lambs on pasture is a tricky operation. Never try to run down a four-day-old lamb. I try to get everything tagged before they are 24 hours old. As I quietly approach the lambing bed the ewe will get up and move off. I will be sure to catch both lambs because if she had one with her when she moves off she may not come back. Once I have all her lambs secured, she will usually turn to face me, stamping her feet. I tag the lambs back away from the lamb bed and then, re-lease the lambs. If all goes well the ewe and lambs meet in the middle right over the lamb bed and soon settle back down. If they start to drift away from the lamb bed I move in a big circle until my presence sends them back where they came from. Remember? I said ear-tagging is a time consuming process.

When I first took a serious interest in pasture lambing, all the pros were advocating "parking" the flock until lambing season is over. This method may work for small flocks, but I have found it quite impractical for big flocks. In my area, set stocking 200 ewes in the spring would require at least 100 acres. 400 ewes would require at least 200 acres. That's a lot of land for a lambing flock to be moving around on, not to mention the job my dogs would have deterring coyotes, foxes, and eagles. I usually move the sheep daily in the spring until lambing begins, then we switch to moving them every two to three days.

As our flock size grew over the years, we began to de-velop a "drift lambing" model. If the flock commences lambing in paddock A, when they need to be moved I set up paddock B and then raise the polywire break fence between A and B to the top of the polyposts all the way from one end of the field to the other. Thus the sheep can enter paddock B at their own leisure and without congregating at a gate.

Any new pairs that choose to stay on their lambing bed are left alone. I intentionally leave enough forage that they don't need to move forward if they don't want to. When pad-dock B has been grazed sufficiently, I set up paddock C and raise the wire between B and C. Again, the flock drifts forward at their own leisure, I don't call them. Once the flock has drift-ed into paddock C, I go back to paddock A and gently nudge any remaining ewes into paddock B. These lambs are now at least two days old. Then I put the polywire fence back down on the normal clips to keep the sheep from backing up into paddock A.

By quietly drifting the flock forward, new pairs have the area to themselves, reducing the risk of mixups. By leaving the polywire clipped to the top of the polyposts, guard dogs have easy access to both paddocks. By allowing the sheep to move before they get hungry, you avoid having a mad rush for the new pasture.

If we get caught with a late spring, and hay supplementation is necessary while lambing, the hay is scattered out in the new paddock before the sheep are let in. Absolutely avoid anything that causes the sheep to bunch up.

For example, if they run out of salt and you show up with that necessary element, the sheep will come running from the four winds to get their share. I will never forget how sick I felt the time I saw that happen, 300 head all pell-mell, humble jumble trying to get to the middle of the mob, sheep piled up two layers deep. Even the ewes that had just lambed came running, not wanting to miss anything that might have drawn such an assembly! Talk about noisy! It took several hours for the ewes to sort out all their lambs and spread back out on the pasture. All that stress could have been avoided by replenishing the salt supply three days earlier. If the flock does run out of salt, go ahead and give it to them, let them fill up and then they can sort out their lambs on their own. I have met shepherds who tried hand feeding grain in lambing season with similar results, don't do it.

To recap:
* Allow the sheep to move before they get hungry.
* Don't force new pairs to move.
* Allow guardian dogs easy access to all sheep.
* Offer tree or brush access as possible.
* Avoid any action that would make the sheep bunch up.
* Salt is cheep. Make it available at all times.
* Always stay calm.

If you want to know which ewes lack maternal instinct, just wait until weaning time and cull any ewe without a bag. You might end up culling some innocent ewes, but you will definitely get rid of any ewes that didn't mother up.

Chapter 18
All Necessary Traits Work Together

All traits that support and sustain reproductive cycles harmonize with each other. After all, reproduction is the ultimate goal of all life.

In contrast, single trait selection may lead to reproductive failure or a shortened productive life span. Following are two examples of how single trait selection can short circuit reproduction.

1. The production of sex hormones slows and eventually stops all bone growth. The the subfertile animal has greater growth potential than its fertile counterpart and will continue to grow while its fertile counterpart moves ahead with its productive life. Selecting for maximum growth without regard to other traits generally results in lower reproductive capacity.

2. Selecting for maximum milk production without regard to longevity of the bloodline may favor high producing young ewes who fall out of the program early in life. High milk production leaves a ewe vulnerable to mastitis and increases her nutritional requirements. In a real life environment any ewe that fails to meet her nutritional needs is going to fail sooner rather than later.

If I take 10 breeds of sheep, each breed varying in size and body type with all 10 breeds originating from operations that feed only forage and press them into my environment, the limiting factors of any environment will eliminate any ewe that is not adaptable to her new surroundings and forage options.

Due to variations within breeds there is a good chance some sheep will be found in all 10 breeds that are adaptable to my environment. The ones that are not adaptable will die from bacteria, viruses or parasites unless I recognize the symptoms of failure and cull them, or prop them up with pharmaceuticals.

Since every environment has its limiting factors, selection must always be capped at optimum, not maximum performance. Some extremes can be overcome with human intervention and indeed that is what shepherding is all about; however, the better a ewe is adapted to her environment the longer she will stay in production and the lower your cost of production will be.

Back to the 10 breeds. The survivors who proved their adaptability will all be fairly uniform in size and phenotype, not cookie cutter but similar. Likewise, reproductive capacity will also be similar across the whole amalgamated flock.

The equalizing process runs something like this:

Forage is fuel.

Rumen is the gas tank.

Endocrine gland systems is the engine.

The sheep's body is the housing for the engine and allocates the energy produced.

In this analogy every sheep will fail without quality fuel or if the fuel is contaminated.

A small rumen (gas tank) will limit energy production. The endocrine system (being the engine) must be fully functional. It is responsible for hormonal balance, vitamin synthesis, reproduction, virtually every function of the body. In fact, the gland system even provides taste and palatability to the meat.

The sheep's body being the housing for the engine is responsible for energy allocation. Huge, overweight, heavy muscled ewes find themselves at a disadvantage. Too much energy is needed for daily maintenance.

Indeed, the sheep's body is an important part of this whole. The endocrine system does not grow in proportion to the body. As the body size crosses optimum the endocrine system quits growing. Now you have a big sheep with a little engine. The result is a ewe that struggles to survive, let alone reproduce.

When a ewe allocates too much energy to any one function, be it milk, fiber or maintenance, all other functions are going to suffer.

All traits that are necessary for continuity of the ovine species work together. In other words, any trait that supports reproductive cycles does not detract from any other trait that also supports reproductive continuity.

Here is a basic list of functions that support continuity of the ovine species.

Immunity.

Fertility.

Soundness.

Easy fleshing.

Lambing ease.

Maternal instinct.

Optimum, not maximum milk.

Unfortunately, some of these traits can be overdone. They each need to harmonize with your context. Just because fertility might allow you to produce a 300% lamb crop doesn't mean your forage will sustain that level of production.

Maternal instinct is self qualifying. A ewe should know how to birth her own lambs and mother-up on her own. A ewe with sub-par instincts as a one- to three-year-old will often become a lamb thief later in life. These halfwit lamb thieves will begin to steal lambs sometimes several days before they themselves are due to give birth. Thieves are quite vocal about what they want, and once they've set their heart on a certain lamb there is no changing their mind.

Speaking of thieves, I remember one ewe who stole five lambs. Once they had nursed the thief their own moms wouldn't take them back. To make matters worse, when old Gerty delivered her own two lambs she didn't even want those. To conclude the story, Gerty headed to the sale barn accompanied by her daughters. I used to think lamb thieves had excessive maternal instinct, but we have come to realize that their instinct is sub-par and they are too lazy to lay down and have their own lambs. Maternal instinct is self qualifying and has nothing to do with stealing lambs from other ewes. Hint: thieves make good lamb burger.

Harmonious selection tools:

Any time sheep are being compared they need to be on a similar diet, of a similar age, and from a similar environment/background. Comparisons become even more relevant if the sheep are on the same farm and have been there for three generations.

Weaning weight percent. This is a tool to compare ewe performance across a group of ewes of varying size. Here is how it works. If a ewe weighs 110 pounds and weans 2 lambs X 55 pounds each she weaned 100% of her body weight. A 150-pound ewe that weans 2 lambs X 70 pounds each only weaned 93% of her own weight. The higher the weaning weight percent, the better. This exposes which ewes can do the most for the least. A ewe should be able to wean 100% of her own weight.

Early sexual maturity. A ewe lamb with a fully functional gland system eating good forage should breed at six to seven months of age to lamb as a yearling. Failure to reproduce as a yearling can be caused by any number of things, but one thing is sure if 90% of the replacements reproduce themselves most of them will probably still be in the flock in seven to eight years. Most of the 10% who failed will probably disappoint you in some other way.

When yearling reproduction falls below 50% it's time to take a look at the overall production strategy.

Chapter 19
Selecting and Developing Replacement Ewes

"Selection is the mightiest tool in the hands of the shepherd; it is the primary means to bring about improvements and to mesh one's sheep with the environment they inhabit." *The Future*

If you wish to develop a functional flock, selection is the key. Assuming you already have a top notch flock, selection will be necessary to keep it that way.

Selection should be tailored to favor reproduction at an early age, adaption to your environment and longevity, which is best described as the duration of productive life.

Dropping all prejudices about leg length, thickness, ear carriage and other cosmetic points, a sheep that is adapted to your environment and lives a long productive life is a keeper. Likewise, her offspring will generally satisfy as replacements.

If you are buying ewe lambs or need to retain ewe lambs that haven't been ear tagged, connecting the candidate with her mom may not be possible, but don't panic. If a lamb is living well post weaning, she will usually produce well. Her ability to thrive on the prevailing landscape is accurately reflected by her hair/wool and hooves.

A healthy happy lamb's hooves will have smooth outer walls. The hair coat should be full color. If she is white, let her be bright white. Likewise if she is black, let her be very black. Black hair shouldn't have brown tips. Dull dead hair hints at a dysfunctional gland system and general debilitation.

If the pedigree of a lamb is unknown, then compare her

to the rest of the flock. Avoid extremes, big, little, thick, narrow, excess facial wool, long legs, short stubby legs, etc. If she stands out from her peers and her pedigree is unknown, discard her. Nature doesn't like extremes. Big lambs are probably singles. Small lambs are often triplets.

Observing your flock shouldn't be like going to the zoo where inmates have very little resemblance to each other. It is far easier to manage the nutritional plane of your flock if they are somewhat uniform.

Don't be afraid to keep plenty of replacements. This will allow you to see how they perform post weaning and then you can sell the laggards later on. A lamb is a lamb until it is a year old. If a lamb is weaned at four months of age that leaves an eight month trial period and any disappointments can be marketed as a lamb in this period of time.

Developing the replacements you select can and should be pretty simple as well. For starters we don't wean our ewe lambs. They stay with their mom and she weans them herself. This eliminates a lot of potential stress and the lambs learn first hand how to function in your environment. No one makes a better teacher than their own mom.

If you do wean your replacements, wait until they are four months old. Early weaning lambs compromises potential for the rest of their lives. The only reason ewe lambs ever need to be manually weaned is if they or their moms must be sold or relocated.

Some folks say they wean so that the ewe lambs can be supplemented to improve development, but I find that a little milk and the low stress atmosphere with a familiar flock will produce results in young lambs unmatched by any supplemental feeding.

Generally, ewes will begin to dry off when their lambs are four months old and most ewes will be totally dry by the time their lambs are five months of age. However, some ewes will continue to lactate for as long as six months, albeit production is minimal after the first 120 days. If a ewe is only lambing once a year it's not going to hurt her to feed her lambs for four months or longer. She will dry off in her own good time and her lambs are the winners for the extended lactation.

Like all classes of sheep replacement ewe lambs will perform best when eating early succession plants such as forbs and annual grasses. Thin ewes and replacement ewe lambs will both gain weight rapidly during early fall.

Good forage and cool weather gets the flock in shape for breeding season. A traditional grain flush is unnecessary. The flock will flush themselves on good pasture.

We always expose all our ewe lambs to the rams when they are approximately seven months old, which means they will lamb at one year of age. The lambs that are physically developed enough to breed do. The ones that are underdeveloped don't breed and/or reabsorb the embryo.

If most of your yearling replacements lamb at 12 months congratulations. You must have a good forage base. Consider marketing the few replacements that failed to lamb. They are probably late maturing and will never reach optimum production.

If 50% or more of your yearling replacements fail to lamb at one year you may want to increase forage selection for your flock during the summer and in the winter. feed more and better hay.

If ewe lambs are well developed pre-breeding and forage quality declines rapidly post-breeding, a lot of bred ewe lambs may turn up open in the spring. The key to sustaining pregnancy is quality forage. If it is not available in the pasture, supplement with small quantities of high quality hay.

Don't expect a lot of growth on stockpiled forages or coarse grass hay. Dormant forage pretty much equals dormant growth/weight gain. Your opportunity to grow your lambs is when your forages are growing and perhaps for 30 days after they begin to go dormant.

Producing one lamb per yearling ewe is my goal. Yearlings that birth twins will often have triplets or quads later in life, which I find undesirable in a forage-based pasture lambing operation.

A 100% lambing rate for yearlings and a 200% lambing rate for mature ewes is ideal. After all, a yearling is still trying to grow while feeding her first lamb and a mature ewe has only two teats, so why give her three or four lambs?

Some folks argue for postponing breeding any replacements until they are 12 to 18 months old, claiming they will develop to a greater size thus becoming a better ewe. But I've never understood their logic. We are after production and profit. The sooner a ewe begins to produce the sooner we start making money. Besides we have had less incidents of difficult births with yearlings than mature ewes.

Lambing at one year eliminates the risk of fatty udders. Plus, yearling ewe lambs can practice playing mother for one lamb rather than waiting until she is two years old and finds herself overwhelmed with twins for the first time.

The bottom line, exposing ewe lambs right along with the rest of the flock is cash positive and allows you to find out sooner what kind of employees you have working for you.

Chapter 20
Rams — The Other Half of the Flock

This important sector of the ovine species deserves a chapter all its own. It is up to the ram to bring about flock-wide improvement by propagating the genetics of long lived productive ewes. And again it is the male lamb crop that furnishes most of the lamb meat in the marketplace.

Apparently to optimize reproduction, nature has the cards stacked against rams. For example, rams are more prone to coccidiosis than ewes. Ram lambs suffering severe vitamin A deficiency early in life never develop testicles and are permanently useless. Ewe lambs suffering equally severe vitamin A deficiency early in life can go on to develop a fully functional reproductive system and bear lambs.

Disease, parasites and reproductive failure are nature's ways to reduce stocking rates when a landscape can no longer support its guests. Kind of a cold hearted host, eh, well maybe. Continuity is a part of regenerative, natural existence. Continuity of the ovine species requires less rams than ewes.

One ram can sire hundreds of lambs in a few year's time. One ewe may produce twenty lambs during her whole life. With continual pressure on the male sector of the population, only the best remain alive and reproductively viable. The net result is more forage available for the remaining flock.

Left to themselves, rams can get torn up and taken out in all sorts of ways. With the mating season naturally occurring at the beginning of winter that leaves no time to build back body condition. With lower disease resilience and natural fighting instincts, rams certainly have a lot of opportunity to prove their right to reproduction. With good management and controlled stocking rates, environmental pressure is reduced to the point that most ram lambs become reproductively viable. Under these controlled conditions it is up to the shepherd to select the rams that ultimately become half the flock.

I take ram selection very seriously. This is our opportunity to mesh the future flock with our natural resources and business objects.

The three traits to focus on:

Reproduction is my first interest. This vital function is the basis of continuity and profit. Yearling ewes can and should produce and rear a single lamb. Mature ewes should birth and rear twins. Selecting for larger litters of lambs is not sustainable on pasture.

Adaptability is my second interest. A sheep must be adaptable to my environment and what my landscape produces, albeit supplemented by salt. Adaptability is the key to controlling costs.

To prove a sheep's adaptability to your environment take away the crutches. The ones that thrive are the keepers.

Longevity, duration of productive life span, is my third criterion. A long productive low maintenance life reduces ewe depreciation costs and ultimately allows me to market ewe lambs that would otherwise be needed as replacements.

Any sheep that remains unsound will self destruct and if a ewe fails to reproduce annually she should be culled. Thus longevity is the culmination of all good necessary traits.

A ram's merit should be based primarily on his mom's expression of these three traits. Of course, he needs to be healthy, sound, well developed and exhibit normal breed characteristics.

Appearance.

A ram that is reproductively sound will have two normally developed testicles, well descended in the scrotal sack. Both testicles should be identical in shape and size. Poll hair on a ram should lay close to the skull. Long silky hair standing up as if electrified signals problems and likely sub-fertility.

As a ram ages his front end will continue to grow after the rear end quits. This is a normal positive characteristic of masculinity, big shoulders, thick neck and taller at the shoulder than the hip giving the appearance of walking uphill.

Masculine rams can make feminine ewe lambs and support uniform gestation lengths culminating in moderate birth weights of offspring. Feminine rams tend to produce the opposite effect.

Capability.

At seven months of age a ram lamb can settle 20 ewes in 18 days. A mature ram can settle 60 ewes in 18 days; however, the ram/ewe ratio suffers as ewe numbers in the flock increase. In other words, if a ram lamb can cover 20 ewes, three ram lambs may not cover 60 ewes. That's because all three ram lambs will try to cover the first 40 head and potentially wear out.

If using ram lambs I like to release a few into the breeding flock at seven day intervals to insure fresh, keen rams are present throughout the first cycle.

Old cowboy wisdom held that two bulls should never be turned out by themselves because all they would do is fight and not get any cows bred. If one bull was insufficient, use three bulls, that way while two are fighting the third will be busy making calves. I learned the hard way that this concept applies to sheep also. If three rams isn't an option, use a mature ram and a ram lamb as a helper. The little guy will keep the big one on his toes and due to their size difference, there won't be much fighting.

While rams are pretty efficient about their job, it never pays to skimp on ram power. In spite of their tremendous ability I'm not going to cut corners on ram numbers. If you enter the breeding season with minimal ram power and one of your rams gets sick or injured, you are going to be lucky if you find a replacement ram available that also satisfies your criterion. Rams are best sourced or at least reserved far in advance of the breeding season.

Advantages of multiple sires.

I am a proponent of using multiple sires of multiple

ages. This may hamper recording all parentage data but the offspring need to stand on their own merit anyway.

Running multiple sires with varying ages is natural, ensures the presence of fertile males and among other things reduces the risk of bottle necking the gene pool down to what we think is best.

Genetic diversity in a population is good. Diversity is what allows a species to adapt to changing environments and other variables.

Introducing new rams.

When introducing new rams or reuniting old ones that have been separated for breeding season, etc., measures should be taken to limit risk of injury. A lot of good rams have broken their necks by fighting with strange rams.

Introduction should take place in small stout pens. Pen them tight enough so that they can't back up and get a run at each other. As they gradually accept each other's presence the pen can be enlarged by degrees.

When you're sure they no longer feel like doing bloody murder they are ready to go out to pasture. Pre-breeding season, rams will fight each other for hours at a time, but we have never lost a ram fighting in the pasture provided they were already used to each other. I guess they know when to call a truce, provided they are fighting with a friend/pasture mate.

Ram lambs also like to fight but I have never known a resulting casualty. They can be introduced any time anywhere with much less risk than with introducing mature rams. For that reason ram lambs make ideal boosters during the breeding season.

Controlling breeding seasons.

Plan the breeding season or the rams will plan it for you and their written plan reads ASAP. Rams are typically fertile year around, so when they find a ewe that is in season you're going to have lambs five months from that date. While fertility of the ewe flock will vary by season and breed, we have found that our ewes may return to estrus 60 to 75 days after lambing. At that point any rams capable of settling a ewe should be removed unless you wish to accelerate your lambing schedule.

To save on chores and other considerations we generally let the rams run with our ewes from the beginning of the breeding season until 60 days post lambing. When not with the ewes every ram still needs a companion. If you have two or more rams the problem is settled, if only one ram, then select a cull ewe or a wether to keep your ram company. A lonely ram will stress, may get sick and go sterile.

Ram lambs viable at four months.

Ram lambs can reach puberty at four months or 120 days of age. All intact ram lambs should be weaned at this point unless you want to breed the flock here and now. Wether lambs should also be examined to ensure nothing was missed.

We graze our weaned ram lambs and mature rams together. This saves on chores and the newly weaned lambs feel a lot more secure with a few older animals on hand.

Wethers.

For the uninitiated, a wether is a castrated male sheep. Sixty days post lambing when we are removing mature rams is the best time to castrate ram lambs.

Emasculatome is the preferred method here. This method uses a tool called an emasculater, which crushes the spermatic cord causing the testicles to atrophy.

This method of castration is bloodless, eliminating the risk of fly strike. According to *Sheep Science* by William Garfield Kammlade, castration with emasculaters leaves no danger of infection, is lower stress and results in slightly greater growth when compared with banding or castration by knife. My own experience with castrating lambs supports the statement above.

By waiting to castrate until the lambs are six weeks old or older the urethra is better able to grow to its maximum diameter reducing the risk of urinary calculi (water belly) later in life. Castrating after 90 days of age is hard on lambs. In my opinion 60 to 70 days old is ideal.

Wether lambs need not be weaned manually. The ewes will see to that job when the lambs are four to five months old. Supplemented by milk and nourished in the low stress atmosphere of a familiar flock, wether lambs will perform much better not weaned than weaned and supplemented with novel concentrates.

Let's recap this discussion:

* Ram selection should be based on ewe performance.

* The three most important traits are Reproduction, Adaptability and Longevity.

* Rams are half the flock and are the ultimate means by which we can propagate sheep adaptable to our environment.

* Plan the breeding season or the rams will plan it for you.

I am not making any recommendations for or against selecting rams from within your flock versus buying unrelated stock. Line breeding is not for the faint of heart, and works best with big flocks; however, as the ram/ewe ratio approaches 1 to 1, inbreeding levels may be reduced. This can be done with ram lambs and they could be marketed as fat lambs after breeding season.

Trading rams with like-minded producers allows small scale shepherds to capitalize on the rams they purchase. Working together always pays.

Whether breeding or buying rams the age old saying stilll applies, like begets like.

Chapter 21
The Need for Managed Grazing

Abundant high quality pasture cannot be wished into existence. Rather, pasture quality manifests when the relationship between herbage and herbivores becomes symbiotic, mutually beneficial. It is the grazier's job to manage grazing animals in a way that is symbiotic for all involved.

Ancient North American pastures were massaged by ever varying herds of bison, elk, deer, antelope and other herbivores, both ruminant and non-ruminant, commingling, migrating and trampling uneaten overripe forages, rejuvenating the sward. These rhythms of forage management were kept in motion by changing seasons, fear of predators and wildfire.

When an understocked landscape developed, a fuel load wildfire would freshen up the area. Once it rained and the forage began to regrow the grazing hordes would move in and graze the regrowth. These multi-species herds and bands would graze the fresh vegetative forage right up to the old burn line, hugging it as clearly as if it were a fence.

Predators and seasonal migrations kept the bison grouped up and moving, creating rest/recovery periods similar to what you give your hay field between cuttings.

With our modern context, fully mimicking the grazing management that took place in 1492 is impossible, besides we don't like wolves, bears and big cats, not to mention wildfire; however we can use fences to control grazing patterns, graze periods and to offer much needed rest periods for each pasture.

Left to themselves, set stocked ruminants and other herbivores will overgraze some areas and undergraze other areas in the same pasture. In both cases the result is soil degradation.

Set stocked animals also develop bad habits such as lounging and pooping in the same places, walking the same path to water or shade creating gullies and selectively grazing their favorite species into extinction. Once their favorite species is history they move on to the next best option. The result is wholesale loss of diversity. If this short sighted practice continues animal performance eventually falls flat.

Restoring predators in the form of electric fence gives graziers an economical way to develop a symbiotic relationship between forages and herbivores.

Grazing management can heal the soil, promote plant species diversity, increase production and sustain animal performance from day to day and year to year.

Why is it that farmers will go to great lengths to make good hay and a lot of it, but when it comes to pasture every management action seems destined to reduce both quality and quantity? The word "pasture" comes from the Latin word *pascere* meaning to feed, and on some farms the word does indicate feed aplenty, but too often it means half-starved animals frantically overgrazing trying to survive. Folks, that is just a desert in the making!

Another critical aspect of pasture management is stocking rate. No place on earth grows forage at a consistent rate year around. We all have seasonal peaks and slumps, good years and poor years. The only way to match forage demand with your forage growth curve is to have a variable stocking rate.

Timing lambing season to begin with the onset of a growth curve helps. Bringing in a herd or flock of disposable animals for the peak growth period is another option. Dropping some paddocks for hay production is also a worthy consideration.

If your permanent stocking rate is geared towards keeping up with forage growth during peak production you will find yourself feeding hay for many months during periods of reduced growth and dormant seasons.

Simply beginning to control or limit your livestock's grazing opportunities does not of itself heal the land or improve performance. The land will only regenerate if the forage/herbivore relationship becomes symbiotic.

Some of the happiest livestock I have seen were moved twice each day and some of the happiest animals I have seen were set stocked. On the flip side, some of the sickest, most debilitated livestock I have seen were also moved twice a day. Indeed there is no magic in paddock rotation of or by itself.

In fact, set stocked animals will usually outperform even well managed stock until forage quality and diversity declines from selective grazing. Set stocking offers low competition, maximum selection and consistency in forage, water and shelter.

The purpose of managed grazing is to:

1. Heal the land, increase production and promote diversity.

2. Limit daily forage access. You wouldn't turn your flock in on the whole hay stack at once, so why the pasture?

3. Done right, managed grazing can sustain animal performance from day to day and year to year.

Peak quality pasture can have a similar feed value to a grain meal mix, and among other things is far more natural to the diet for a ruminant. Indeed, it is well documented that animals can gain weight on pasture just as fast as in a feedlot. The key is forage quality, and management is the key to growing quality forage.

Chapter 22
Pasture Management — Best Practices

Why all this fuss about pasture? Because if you are going to be a shepherd you will have to feed your sheep, and pasture is the best way to feed the flock.

For years I knew that sheep grazing high quality diverse forages outgrew and out performed sheep that were confined to scientifically formulated TMR (total mixed ration, often with a major grain component). Yet I laughed at the notion that forage could replace grain on a pound for pound basis.

Then one day I overheard a feed salesman/nutritionist tell a fellow shepherd that his fourth cutting alfalfa hay had more feed value per pound than a 16% protein lamb grower/ grain ration. This came from a feed salesman! In other words my friend could pay more per ton for good alfalfa hay than for grain!

That incident set me to studying printouts showing protein and TDN levels of common forage. Not surprisingly the analyses showed better readings on pasture than hay, but both cases varied by maturity and species. And yes, these forage tests agreed with the feed salesman.

Most forages pass through a stage that is similar to common grain rations in terms of TDN and protein levels. Acknowledging this truth of pasture potential brings up the question, "Why don't we value and utilize our pasture more than we do?"

The first step is to acknowledge the potential. The second step is to learn how to manage your forage to develop and maintain quality forages. And the third step is to actively manage your forages.

Respect, understanding, performance.

Those three words would transform many marginal crop fields back to pasture! May that day come quickly.

I promise not to let this chapter get too lengthy. There are plenty of good books on pasture management for further consideration, however forage is and has been the foundation of my sheep business. This book wouldn't be complete without a discussion out in the pasture. So, let's pretend we are going on a pasture walk, taking time for observation.

What species are the sheep eating?

How much plant residual is left?

Do you see any bare soil?

What about manure quality?

Are the sheep grazing or ruminating?

The best way to determine the maturity level of your grass is by counting the leaves. Once a grass stem has two leaves the plant is entering a peak growth phase. Peak growth will continue through the development of leaf number four and then begin to slow with number five. By the time leaf number six is formed, leaf number one is dying.

Livestock usually make the greatest gains when forage is harvested in its peak growth phase. This pretty well holds true whether the grass is actively growing or its growth is suspended in dormancy or curing.

Here is a good rule of thumb. You can begin grazing when your grass stems have developed more than two leaves, provided two leaves or the equivalent are left intact. Regrowth will not be slowed significantly. Said another way, if the grass has three leaves, graze one. If it has four leaves, graze two. Reducing the plant solar collector to less than two leaves may temporarily reduce plant growth by as much as 80%.

A paddock should not be regrazed until all the blades of grass have a sharp point. If the flock is returning from the grazing circuit to paddocks that contain grass leaf tips that appear mowed off the rest period is too short. You may want to consider reducing animal numbers or increasing the acres in your grazing pool.

Maturity of legumes is best determined by bloom stage. Alfalfa is ideal when 10% of the plants are blooming. I like to graze red clover after it has reached full bloom and 10% of the blooms have turned brown, which results in a fresh seeding of clover every time a paddock is grazed.

Forbs vary tremendously in their need for a recovery period. Many of them cycle in and out of a palatable state. If the flock is not eating a certain plant just wait. It may well make it onto the menu when it becomes more mature.

Grazing pastures composed of a variety of forages will result in some species being under ripe, some perfect and some plants will be over-ripe. Just do the best you can to favor the species you desire.

Leaf counts/maturity correlations are pretty uniform across most species of grass. On the other hand plant height at the point of maturity will vary with fertility, moisture, species and other conditions.

It is NOT the height of the grass plant that determines when the sward is ready to graze, but the maturity level of the sward. When grass is fully recovered and quite mature it should be grazed, even if it is only three inches tall. Grazing stimulates regrowth and optimizes soil regeneration. A plant that has been grazed will pump much more carbon into the soil than a plant that has matured and is dormant.

Forcing livestock to graze undesirable plants will lead to overgrazing of desirable plants. In that scenario your management would favor the species you don't want. Mowing or bushhogging would be a better way to reduce competition from unwanted herbage.

A pasture's composition will change somewhat seasonally. The pasture sward should be eaten, mowed or trampled to the ground once a year to allow all species an equal opportunity to grow. Otherwise continual selective grazing will result in species loss. Likewise a full recovery at least once each year will allow maximum root development and seed production.

As a grazier you can control the species mixture of your pasture to a large extent by keeping your grazing management flexible and adaptive. The species composition can be shifted in a major way in two to three years. The outcome may be negative or positive, depending on your management.

Management practices need to be adapted to favor positive outcomes from each pasture, season and year. The context is always changing so staying adaptive and flexible to favor what you want is necessary. Be slow to kill something that wants to live and don't try to keep something alive that wants to die!

Non-selective grazing is best viewed as a bite taken out of everything, not everything grazed to one even height. Trying to graze everything down to one even height will end approximately at ground level. However, leveling rank stands of grass will promote forage diversity because forbs and legumes will have an equal chance to grow in the weakened grass stand.

A common mistake cattlemen often make when getting sheep or goats to control weeds and brush. Grazing management is tailored to kill the weeds. Once killed your sheep just lost their food source. You don't keep cows to kill grass. We only expect cows to prune the grass, raise calves and prune more grass. This view of weeds should be adopted by shepherds. Utilize the weeds, raise lambs and graze more weeds.

In grazing circles the term "trigger height" suggests that when the overall sward has been grazed down to a certain height a paddock shift is triggered. I suggest that shepherds should add trigger species to their monitor.

The trigger species are the first three to five plant species that the majority of the flock eat right after they are moved into a new paddock. When those three to five plant species are grazed out a paddock shift should be triggered, provided you wish to optimize sheep performance.

Keep tabs on your forage inventory. The simplest math is to divide the acre units grazed by the flock each day into the total number of acres at your disposal. This calculation gives you the number of days the flock may continue to graze.

If the flock is grazing 50% of the biomass and the forage doesn't grow any more after that day you can probably make a second round on the pasture taking another 50% of the remaining residual. So if your first round lasts 70 days, then the round following should last 35 days, unless of course if the forage continues to grow.

Being able to measure your forage bank out in front of you is peace of mind, just like knowing how many bales of hay are in the barn.

Another method used to measure forage inventory is to physically measure your forage depth on an acre by acre basis. A healthy diverse stand of forage may produce 250 to 300 pounds of dry matter per acre inch of standing feed. So, if you have eight inches of dense forage it should yield 2000 pounds of dry matter/acre. Let's leave 500 pounds/acre as soil cover for an earthworm party. That leaves us with 1500 pounds of dry matter/acre. Sheep will eat approximately the equivalent of 4% of their body weight in dry matter daily. So if your ewes weigh 125 pounds each thus needing five pounds of dry matter daily, then your 1500 pounds of useable forage/acre will generate approximately 300 sheep days per acre.

Limiting access to stockpiled forage will stretch your resources and sustain performance further into the dormant season. Managed grazing has been referred to as controlled starvation. While rationing forage to dry ewes makes economic sense no sheep should be reduced to a starvation plan. Moral: ration but don't starve.

Is fertilizer positive or negative? I am going to exercise my freedom of speech and state that it is often a negative management practice. Here is why.

Nitrogen (N) fertilized grass can cause a magnesium deficiency in livestock. While this is a common disorder occurring in many flocks and herds we have never had an issue with our own livestock, simply because we have never applied N fertilizer.

Nitrogen fertilizer also reduces the presence of legumes in the sward. The same thing happens if excessive quantities of hay are fed on a pasture. In both cases after a few years the excess Nitrogen will dissipate or leach out of the soil and with sufficient animal impact the legumes will reappear. Fertilizer has also been proven to increase the toxicity of fescue. All in all, dependence on fertilizer has a propensity to leave your pasture operation bleeding red ink — for a variety of reasons the least of which is not the cost of this volatile input.

Thank you for taking time to walk through the pasture. One pasture walk won't make you a professional manager, but you will never be a pro unless you take time for that first walk.

Chapter 23
Naturally Occurring Forage Chains

When I think of a forage chain, I like to think of each link of a chain as a complete loop of nutrition, a community of plants with each filling its place. When one member is inactive, the rest of the plant community steps up to fill the void. Nature always provides something as long as we control stocking rates and leave the plow, bush hog and herbicide in the shed.

Every plant on earth has a beneficial purpose. Some "weeds" just get misused and a lot of other weeds are not utilized at all. By adding more species of grazers we can begin to use more of the forage species God has given us. Each species of herbivore has its own preferred forages and does best when allowed to consume a diet suited to its needs.

Each plant species has its own mineral profile and given enough forage diversity, livestock can go a long way in balancing their mineral needs. If your sheep are not eating a certain weed do not panic, change your management, or just be patient, that weed may end up on the menu yet. If we focus on grazing the desirable species in a way that favors those plants, they will generally crowd out the less appreciated weeds over time.

Each plant species has a unique place to fill in the forage chain. The more species we have the stronger the chain becomes. Species multiplication won't happen just because we throw more seed out on the pasture. Species diversification is best accomplished through livestock impact; but this discussion is going to center around the use of our available weeds, assuming you already have a lot of diversity.

Each plant has its own ideal time to nourish your livestock. Here in North Central Missouri the forage chain may begin in the early spring with fescue, yellow dock, and dandelion, to name a few. Dock and dandelion both cleanse the bloodstream and are a great liver tonic.

As the weather warms and we pass last frost, fescue gets replaced by other grasses growing in the same pasture. At this point clovers, compass plant, black eyed Susan, asters and chicory help to form a loop of the chain. Trees and bushes have leafed by now and provide their own special nutrients.

Did you know chicory and all blue flowering plants are high in copper, and copper delivered in plant life is the ultimate way to supplement?

As we progress into summer the weed options continue to change: ragweed, horseweeds and ironweed come and go. Ragweed and horseweed are best before pollinating. If these weeds pollinate before they are grazed, let them go. Livestock like ragweed seed once it is mature. Ironweed needs to be in full bloom before it's attractive. Excessive legumes can cause metabolic disturbances and physical disorders. Allowing legumes to fully mature is beneficial.

The grass and legume component will continue to shift also. Crabgrass and foxtail grow well in July and August when the cool-season grasses are semi-dormant. Korean lespedeza can be a major forage source in August and September in a dry year. Korean Lespedeza is a non-bloating legume.

As frosts resume in the fall, fescue becomes valuable again, as long as it's not mature. Fescue grows best in cool weather. No longer stressed by heat, the plants grow stronger and less endophyte infected. The endophyte is just a plant parasite attacking a heat stressed grass plant. Honestly, fescue has no place in Missouri but since it has become a part of the local forage base I utilize it in the forage chain. This means during the cool months of the year.

Have you noticed those marble sized, yellow berries hanging on horse nettle plants? Sheep and goats love these little fruits. They get ripe around first frost in the fall and hang on the plant all winter. These berries provide a durable form of carotene. Carotene is the ruminants' basis for vitamin A synthesis.

As we move into winter, the forage choices narrow a little but that is okay, assuming the birthing season was when it should have been. Maintaining adapted livestock on frozen grass is a breeze. Trying to get stock to birth and raise strong babies on stockpile is a nightmare. Match livestock peak nutritional demands to peak nutrient availability. Your stock is the beneficiary or victim of your management and available forage.

It is possible to have some diversity in the winter diet. My goats love the red berries on buckbrush plants, especially in late winter. Again, a source of carotene. Cows, sheep, goats and pigs all benefit from acorns. Ragweed that was allowed to seed out in summer is now sticking above the snow. Judging from the way stock clean these stems off it must be pretty nutritious.

Rosehips are a sheep and goat delicacy. Rosehips are high in vitamin C and who doesn't need a little extra vitamin C in the winter? Eastern red cedar is a sheep and goat favorite January through March. Red cedar stimulates the immune system and increases circulation.

Notice I did not say:

*put your cows on a fescue field in winter,

*put sheep on a dandelion dock field in spring,

*put sheep and cows on a crabgrass-foxtail field in July-August.

An effective forage chain may contain 100+ species in a year and all these species can grow on the same acre of land. As the spring dominant forages dry up, the summer forages take over and then give way to fall and winter. I find this kind of organized chaos to be much simpler than planting 100 separate fields into monocultures to capture the benefits of 100 forage types. Besides, all grazers benefit from a wide variety of forages in the diet every day. Diversity is key to health, reproduction and parasite resistance.

Just don't try to make your stock eat all the diversity at one time. Each forage specie has its own window when it will be the most beneficial. Intensively manage (monitor) what the stock are eating. Adapt to circumstances, some paddocks will have more forage species on the menu. Paddock size will need to vary. You may need to mix up the rotation to take advantage of some prime weeds. Do not let it bother you when the stock MOB down the species that are currently undesired, "think fescue midsummer."

What could be more exciting than a huge flower garden with 100+ species of grasses, legumes, forbs, shrubs and trees? With something always flowering or fruiting, producing a wide range of minerals, vitamins and immune enhancers, and multiple species of healthy livestock will be content to harvest the bounty.

Chapter 24
Don't Force Rapid Diet Change

Sheep are characterized by abrupt changes in behavior and food choice, but in spite of their ability to change on a whim they may not adjust well if a diet change is forced on them rapidly.

Any time a radical change is necessary it is desirable to offer access to the previous food source the flock was used to for a few days until the rumen bugs adapt to the new food.

Transitioning from hay to grass is best done on a sunny afternoon when the forage is dry. Continuing to offer dry hay as long as the sheep will eat it, even in small quantities will help to stabilize the rumen. If sheep need it they will eat it. For years I was told that low quality hay or straw works best for this transition period, the main purpose being to slow the passage of short lush grass through the digestive track.

It is true that a desperate animal will eat some pretty raunchy hay, but they will eat a lot more of it if it is good quality. In fact I save the best hay for this transition period. Here is why.

Short, wet grass on a cloudy day in spring is a very low energy food source. Good, high energy hay will reduce the passage rate allowing better nutrient assimilation and will boost energy levels in the diet. This transition period may only last three to seven days but some really good hay fed in small quantities can make a huge difference in animal performance the rest of the year. If they don't eat it they don't need it. Don't force consumption.

Transitioning from grass to legumes is also a challenge.

The ideal pasture will have a broad mix of forages including several species of legumes; however some farms are characterized by pastures that are free of legumes and hay fields containing only legumes.

If the sheep have had no recent exposure to legumes, turning a hungry flock into a wet stand of legumes can cause indigestion and bloat.

Preventing bloat is a two step process.

1. In the morning move the flock onto a fresh paddock similar to what they have been grazing. Let them fill up on non-bloating forage.

2. Once all the dew is dried off in the late afternoon, turn the no-longer-hungry flock into the legume field. Having their bellies full of non-bloating forages they won't be able to eat enough alfalfa or clover to cause bloat. But if you don't want to take a chance then remove the flock just before dark and repeat steps 1. and 2. for two or three days until the rumen bugs have adapted to the new forage.

If your regular pastures include a healthy mix of clover or alfalfa, transitioning to a monoculture is not a big deal.

Birdsfoot trefoil, lespedeza, etc. are non-bloating legumes posing no threat and adding diversity to the diet. Diversity equals resilience for all involved. Monocultures are high maintenance, high risk and unnatural.

Perhaps the highest risk in change of diet is from forage to grain. This requires a massive change in the microbiome. The change starts to occur when grain consumption exceeds .5% of the sheep's body weight or ½ pound of grain per 100 pounds of sheep.

Increasing the grain portion of the ration faster than .5% of the body weight per week can lead to overeating disease *Enterotoxemia*, a highly fatal disorder. Treatment is basically useless; prevention is the answer.

High levels of grain consumption, regardless of how slowly the ration was increased, can cause a condition known as acidosis, excessive stomach acid. Acidosis predisposes animals to low immunity and ill thrift. The answer is to keep enough hay in the diet to buffer the ph. After all, legume hay can have a feed value equal to or greater than standard grain finishing rations and doesn't present all the inherent risks associated with grain.

No animal being retained for reproduction should ever consume grain in excess of .5% of its body weight. Zero grain would generally be better.

Grain is one good horrible example of a food source where sheep will fail to self limit consumption without fatal consequences. In other words the saga "sheep know best" does not hold true when sheep are presented with an unnatural opportunity to eat grain.

Grain in its native state is long on fodder and short on seed. Plus, if ruminants had access to wild grain at all they would have had access throughout the growing season and likely would have grazed the plants, thereby reducing grain production and subsequent consumption.

Properly grown plant communities with a lot of diversity generally pose no risk to adapted sheep. If the flock has been targeting crabgrass and suddenly chooses to begin eating cocklebur leaves and a week later they are targeting ragweed, that is fine. Rapid change is okay when the flock is self regulated and has had a lot of previous exposure to a forage option because in that context, "sheep know best."

Chapter 25
Free Choice Minerals Can Be Free!

"Rough coats? Respiratory trouble? What's your mineral program?"

"The XYZ blend, I have been feeding it for the past 14 months," I mumbled.

"Oh well, the symptoms you described would suggest an Iodine and Selenium deficiency. Maybe Copper also." This came from a salesman. "My program is high in those elements. My formula also contains the clinically correct amounts of vitamins A, D, and E to support immune function. You want to offer your sheep a ph buffer such as sodium bicarb. We also sell kelp meal. If your sheep are needing additional Iodine, they will have to get it from kelp because my program is legally maxed out on Iodine."

That conversation is a hybrid of many conversations I have had with nutritionists and salesmen over a period of eight years.

I was desperate to find something that I could buy to feed my sheep that would restore and sustain flock performance at historic levels. Every time I thought we had found a silver bullet program we would begin to increase sheep numbers. Or we might lose a leased farm, thereby increasing competition for the remaining weeds. As forage selection fell, sheep performance would too, in spite of the silver bullet mineral program.

After becoming quite dizzy from this roller coaster ride I began to notice a common factor shining through all the variables. A ray of hope. Yes, I was getting discouraged, in fact depressed.

All those friendly salesmen with all their products had failed to restore and sustain my flock's vigor and performance to its historic level.

What is a dumb sheep farmer supposed to do if no one else can solve the mystery?

But ah, that common factor connected with the good years was the light at the end of the tunnel. Sheep performance always rose in response to increased forage diversity provided they also had access to clean water and adequate shelter.

The forage diversity phenomenon held true regardless of which mineral program the flock was on. That startling realization took me back to my boyhood, first with the Gulf Coast Native and later with Katahdins. Kansas rock salt was the only mineral supplement available for my flock and yet performance exceeded anything I had experienced since.

What is it about lowly, despised weeds that keeps a sheep healthy and makes a lamb grow? Why is it that weeds, the enemy of modern farming delivers results like no other mineral program I have tried?

I was told that our soils are depleted and what minerals are present are locked up. Whenever weeds were available to the flock, sheep performance suggested otherwise. Who am I to argue with animal performance, which after all is the key to measuring nutrition?

I timidly dropped our mineral program and started to aggressively propagate weeds in our pastures. Sheep performance rose. Talk about euphoria! Then I remembered two statements I'd heard years earlier:

* Every plant species has a unique mineral composition.

* Weeds are accumulator plants, collecting and storing in their tissues up to several hundred times as much of a trace mineral element as an equal quantity of soil contains.

Wow! No wonder a sheep thinks the best things in life grow in the pasture. If the minerals in our topsoil are leached out or locked up, weeds are going to fix that problem.

As sheep performance continued to rise and my quality of life and thought kept pace, I began to wonder if indeed plant life could serve as a free choice, full spectrum, comprehensive mineral program.

Pushed by curiosity I began counting non-woody plant species on land we own and rent. The tally, 117 species I could identify, approximately 30 species I couldn't identify. Next we counted woody species and came up with 51. That gave us an approximate total of 198 species, 198 forage options.

There is an estimated 400,000 species of plants on earth and with only 92 minerals/trace minerals needed to support life and reproduction. Full spectrum mineralization from herbage was beginning to look quite logical. The difficulty is growing and maintaining enough diversity.

Grass is said to take up the broadest range of minerals, but at much lower concentrations than forbs contain. Grass is better suited to a cow's needs than for sheep that are growing or lactating. However, sheep will do okay on grass during the low ebb of their annual production cycle.

With a sheep's natural production and growth, mineral needs are all over the board. Individuals' needs vary from the flock average and the metabolic ability of each sheep also varies. Sheep are master nutritionists, and due to rapid bio-feedback they can balance their own ration if given enough selection. Broad spectrum mineral blends out of a bag don't allow sheep to fill their needs when the production phase calls for a spike in Phosphorus, Sulfur or Magnesium.

Research has shown that forage preferences can change within a day's time. Personally I have observed sheep eating a forb with great gusto and a year later they won't touch that species. Still later the flock is back to eating the same plant again. Such actions may appear abstract but sheep know best and there must be a cause. What it is I don't care.

As my respect for weeds grew I became curious which species of plants in my area accumulated the elements two-legged nutritionists said my sheep were lacking and whether the sheep showed any preference for those plants. Here is what I found.

Mineral: plant source, partial listing.
Iodine — wild garlic, mushrooms, brassica leaves.
Selenium — chamomile, nettle, yellow dock.
Copper — chicory and other blue flowering plants.
Phosphorus — sunflowers, nettle, legumes.
Calcium — legumes, dandelion, shepherd's purse.

Obviously plants accumulate multiple mineral elements. As it turns out my sheep eat these plants with gusto, but not all the time. Sometimes it's nettle and yellow dock. A month later they may want chicory and legumes, later still sunflower, wild garlic and chamomile.

Given the option sheep will eat many different species in a single day but usually fill up on just a few.

I am told that minerals in plant life (organic form) are approximately 90% bio available, and minerals out of a bag (inorganic form) are approximately 20% bio available. No wonder sheep thrive on minerals processed and packaged in my own pasture.

While it is true that some plants accumulate salt, there is generally a lack of salt in high rainfall environments. Such is definitely the case here in North Central Missouri, and sheep will become salt crazy if deprived for long.

I like to feed Kansas rock salt because it is from an ancient sea bed, has a low Iron content and being somewhat local it's fairly inexpensive. The sheep love it. As with any program individual sheep who fail to thrive should be culled.

After years of crawling through a long dark tunnel, spending tens of thousands on consultants and mineral programs, suffering excruciation death loss and sub-par animal performance I finally stepped back into the bright sunlight to find myself back in the same paddock I left eight years earlier. It was good to be back.

There may be some well formulated mineral programs out there, but I dare to say they all will work a lot better if the livestock you have are already thriving on your available forages.

Chapter 26
Supplements —
Gut Friendly and Otherwise

This is a touchy subject. Everyone likes supplements, right? Helpful salesmen stand ready to advise, the product list appears endless and is still growing.

Why is it that people will drop whatever they are doing to attend a gathering featuring a supplement salesman promoting the newest concoction and afterwards attendees fill up on free pizza and ice cream, meanwhile resolving to spend their gold on this new program?

Why is it that these same folks don't have time to attend a pasture walk and learn how to reduce their dependance on off-farm products? The money saved by independence would buy much more pizza and ice cream than you'll ever eat at those sales shebangs. Plus if you're buying the refreshments, you can choose the flavors.

The key to animal performance is balancing stocking rates with carrying capacity while offering each species of grazer a native diet. Consequently, multi-species operations will achieve a greater level of production per acre than will single species operations.

When our stocking rate for sheep exceeds the pasture's carrying capacity, sheep performance falls. The temptation to sustain performance and ultimately flock numbers by beginning a year around supplement program can be strong.

Giving into the urge to maintain a larger flock than your pasture can support is called egonomics. Conversely, economics, profit driven decisions, will usually suggest destocking.

However, if forage quality is temporarily low, gut friendly supplements used wisely can make a huge difference in animal health and performance. Case in point would be feeding small quantities of legume hay once the protein, energy and carotene levels of stockpiled pasture decline in late winter.

The supplement cost/benefit ratio can help determine if, when and what to feed. Sheep can live, thrive and reproduce without supplements beyond access to salt and hay if the snow is too deep for grazing. Stocking rate is the key.

Here at Still Water if the projected supplementation cost for a ewe for one year exceeds one-sixth of the commercial value of two weaned lambs I take it as a hint to consider reducing sheep numbers.

Feeding hay during the normal grazing season is best described as substitution, not supplementation. Substitutes are expensive and unnatural.

Following is a layman's inventory of supplemental knowledge.

Salt. In some environments livestock won't eat salt due to the quantity of salt contained in forage and water. That is certainly not the case here in North Central Missouri. Sheep and cows will go crazy for salt if deprived for long periods. Goats deprived of salt will binge with fatal consequences when they finally gain access. In that case salt should be rationed for a few days to prevent over consumption.

The phrase worth her salt was probably coined long ago when transportation costs were so high that salt was sometimes used as currency. In that case, a ewe sure would need to be productive to earn her salt.

A legend passed down from the Old West recounts how some ranchers were too cheap to salt their cattle. It was a common practice in those days for ranchers to commingle their steers for the trail drive to the railyards in Kansas. One night a steer belonging to a cheapskate outfit went crazy for salt, snuck into camp and ate one of the wrangler's only pair of jeans (must have been sweaty). A ballad moralizing the event was composed stating that cows should have salt when the grass is turning green and when you go on a trail drive take an extra pair of jeans.

Stock salt should be unrefined, unadulterated and offered free choice at all times. Natural sources of salt from the sea or from an ancient seabed, such as Kansas rock salt contain a natural blend of minerals and trace elements. We avoid any salt that has been adulterated with an anti-caking agent.

We also avoid pink salts due to their high content of Iron. The Iron may be inorganic and unavailable but that doesn't necessarily keep it from wreaking havoc.

Salt blocks limit ingestion. Salt should be offered loose or granular. A grazing animal simply doesn't have time to waste licking a salt block.

Refined, pure white salts have been stripped of their natural trace mineral content and are best used to melt ice on the sidewalk.

Salt summary: stock salt should be unrefined, unadulterated and available free choice at all times. Salt or mineral salt as some call it improves gut health when consumed only as needed. I never force intake.

Vitamin A. Vitamin A is found in animal products — meat, milk, eggs. Ruminants can synthesize vitamin A from its precursor carotene. Carotene is abundant in leafy green forages, yellow coloring matter of corn, carrots, turnips, fruits and berries. Carotene gives nature its colors and in turn gives color and life to the coat — hair, wool, feathers — of living creatures.

The liver is responsible for breaking down carotene and synthesizing vitamin A. Vitamin A can be stored in a healthy liver in quantities great enough to last a sheep three to four months.

Cured forages, whether hay or stockpile lose their carotene content through oxidation. Thus fresh living forage is generally the most potent source. If drought or dormancy reduces standing forage to a brown mass of carbon, after four months of a brown diet it would be good to begin supplementing with good green hay.

If a ewe is deficient in vitamin A when her lambs are

born they will not live beyond nine days.

Pink eye, contracted tendons, difficult births, respiratory issues, abortions and infertility all suggest a vitamin A deficiency and a lack of carotene in the diet. If carotene levels have been high but problems persist, the liver is likely working double time purging out toxins. Likely trouble makers would include moldy hay, ergot, endophyte overload, dirty water, antibiotics.

We have never had noticeable success feeding vitamin A as a supplement. However, my sheep have always responded to high quality carotene rich forages. Vitamin A is best acquired in the natural way.

Vitamin D. Vitamin D is found in forage. Some species contain more than others. However, the chief source is sunlight. Vitamin D3 is synthesized in the skin in response to sunshine. Skin, hair and wool pigmentation, dark colored skin and fiber, block ultraviolet rays reducing vitamin D systhesis. Brown and black sheep will spend more time sunbathing than sheep with white coats and pink skins. Thus, sheep with access to shade and sunshine can regulate vitamin D levels.

Naturally, all vitamins can and should be obtained from nature rather than a bag or bottle.

Minerals. Minerals in the organic form in plant and animal matter are approximately 90% bio available.

Minerals in the inorganic form, mined and sold in a bag, are approximately 20% bio available.

Both forms require digestion, which would you like to consume?

During my first years with sheep, Kansas rock salt was the only mineral supplement I fed. Performance was great, but as sheep numbers rose, forage selection was reduced, animal performance fell.

For years I hunted for a mineral/vitamin program that could sustain the flock's performance in spite of stock rates. Every time I thought we had found the perfect program we would intensify sheep number per acre. Performance always fell. Subsequently when forage selection improved, animal performance kept pace.

We never found a program that would sustain life and reproduction without moderate quality forage suited to a sheep's metabolism, but we did learn a few lessons in the line of gut health.

1. Never feed a mineral product containing enticers such as grain, dry molasses or sugar.

2. Salt/mineral blends may be okay, provided free choice salt is still available at all times. Otherwise the salt component of the mineral blend may entice the sheep to eat minerals they don't need.

3. Most programs are over priced.

I am no longer searching for a miraculous mineral program. Forages deliver the best minerals and we are leveraging our pasture sward to capitalize on the minerals in our soil. The dirt under our feet hasn't been depleted of minerals. Folks have simply narrowed the forage opportunities down to where sheep struggle to balance their mineral needs.

Animal performance is the key to measuring nutrition and our sheep are thriving on our forages supplemented with salt. Bring on diversity. God put forbs here for a reason.

Apple Cider Vinegar (ACV). Raw unpasteurized ACV is a potent source of energy, high in Potassium and full of enzymes. ACV can improve digestion of rough forages by up to 20%.

Sheep will generally self regulate intake. If forage quality is adequate they won't touch the ACV. If the forage is quite lignified they will consume prodigious quantities.

Three pounds of salt mixed with one gallon of ACV will keep it from freezing and takes the bite out of the vinegar. If the flock is binging on vinegar, increasing the salt ratio will reduce consumption. Monitor the manure. If it's runny like a lava spill they may be overdosing.

Sheep and other livestock that are unfamiliar with ACV will get with the program quicker if you add one gallon of wet molasses for every three gallons of ACV.

Depending on price and availability, we offer the sheep

the ACV/salt mixture during February and March to boost energy and improve digestion of stockpiled forage. If we have access to good quality hay the vinegar is not needed.

To put ACV in perspective, it is the most valuable as a supplement for low quality forages. Something needs to change if the flock is consuming vinegar year around. Rotten apples from your own orchard will have a similar positive effect on the sheep's gut.

Natural Sodium Bicarbonate (Baking Soda). Soda is a salt minus the chloride. Being chemically very alkaline it is a powerful quick fix to raise the sheep's ph if they are suffering from acidosis.

Grain. Grain is natural to a sheep's diet but only in very small quantities. After all, grass and weed seeds are grain and yet where in nature do you see vast fields of heavy grain occurring without human cultivation? They just don't exist.

When provided an opportunity to eat unnatural quantities of grain, sheep will binge or overeat. Excess grain consumption leads to all sorts of problems, beginning with the gut and perhaps ending with foundering. Excessive grain feeding at any stage of life may lead to unending hoof growth and regular hoof trimming.

There is no magic in grain. Under the right conditions sheep can fatten just as well on a strict diet of forage. But if grain must be fed, care should be taken not to cause acidosis or any other major rumen disturbance. Limiting grain consumption to .5% of a sheep's body weight is said to not affect a ruminant's digestion. That means a 100 pound sheep could eat half a pound of grain without causing imbalance of its gut.

Over conditioning breeding stock is detrimental to reproduction. Fat deposited in the udder impairs milk production and fat accumulated in the scrotum of the ram may cause sterility. Of course over conditioning can happen with grain or forage. Nature discriminates against obesity just as much as skinny animals.

Legume hay. Depending on species and maturity at harvest and other conditions, the crude protein content of legume hay may range from 16 to 21 percent. Large quantities of hay in the diet is much more natural than grain. Hay being gut friendly, it is the ideal supplement for low quality stockpiled grass in late winter.

If you are paying $400 for a ton of grain (corn/soy blend at 16% protein) you could switch to alfalfa hay, pay more than $400 a ton for the hay and still come out ahead due to a healthier gut. Said another way, good legume hay can replace grain on a pound for pound basis.

Being rich in carotene and highly digestible, legume hay is probably the best way to boost vitamin A levels in late winter or during prolonged drought.

Herbs. Herbs have been gaining popularity as alternative treatments for virus, bacteria and parasite upsets in livestock. When sheep are getting sick from a lack of forage diversity, herbs can be very effective forms of medication without all the side effects of antibiotics.

The amazing thing about herbology is that herbs are simply glorified weeds that we have learned to use. Consequently they are becoming a marketable product, another way to spend farm income. I suggest propagating as many species of weedy herbs in your pasture as you possibly can, thereby allowing your master herbologists (the sheep) to self medicate, thus reducing your supplement bill.

Chapter 27
Plant Communities in Perpetual Transition

The plant communities in our pastures are composed of species that thrive under the prevailing circumstances. With an estimated 400,000 species of plants on earth one or more species will likely be adaptable to each variation of climate, fertility, soil chemistry, soil moisture, soil biology and other variables.

If conditions are favorable for any given species and seed is present in the soil, that species will germinate and proliferate. Seeds can lay dormant for many years until conditions are favorable for growth.

With a shift in moisture, biology or fertility, the dominant plant species will be succeeded by species that thrive under the new conditions.

The general order of succession is weeds to grass to brush to forest, provided environmental conditions allow succession to advance that far.

Under ancient natural conditions as the above ground biomass changed forms, the herbivores attracted to that landscape would also change in response to food availability. If weeds and brush are abundant browsers proliferate. If grass becomes dominant then grazers would proliferate.

As a grazier, my first interest is to choose herbivores that will thrive on the forages my pasture is currently producing; however I digress. The point of this chapter is to impress how we can alter or improve plant succession with grazing management, so here goes.

Bare soil invites weeds and annual grasses, however, the present condition of that soil will affect which species germinate. Seven years ago we cleaned the silt out of the pond behind our house. There was a massive 10 feet of sediment accumulated from the neighbor's property upstream.

For two years this pile of silt grew smartweed six feet tall. Year three the smartweed was noticeably shorter and was interspersed with wild millet. Year four brought millet, giant ragweed, foxtail, wild lettuce and a few small sickly plants of smartweed. Soil conditions were changing. The smartweed that kicked off succession helped to heal the soil and caused its own demise.

Year five brought an increasing number of weeds that thrive in well drained soil, plus perennial grasses including brome and orchardgrass began to colonize. By year six the massive mud pie transitioned to a dominant stand of perennial grasses, yet I never planted one seed on that pile of mud.

The truth is I could have sown that mud pie with perennial grasses the first year, but it still would have only grown smartweed. Likewise there is plenty of smartweed seed in the soil today, but it's not going to grow unless we recreate the anaerobic conditions it likes.

Nature always seeks to heal the land and to cover the soil. If a species of weed, sedge or grass is growing where you don't want it, remember this, it found an opportunity to grow where nothing else would.

I have a 20 acre field that suffered from sheet erosion 30 years ago. When I bought the farm the only plants growing in it were serecea lespedeza and broomsedge, also called poverty grass. There are bare areas in the field that have not healed over yet. Every year the serecea lespedeza and broomsedge creep further into the remaining bare areas. And every year on the opposite side of the field perennial grasses aided by legumes and forbs continue to crowd out the serecea lespedeza and broomsedge.

To be sure I am thankful for serecea lespedeza and broomsedge because they are growing where nothing else will.

In turn, both species will disappear when conditions become favorable for better forages.

If your plant community is drifting towards species that you view as invasive you may want to consider whether your grazing management is detrimental to the species you prefer. Selective grazing will almost always favor the plants that do not get grazed or trampled.

Forcing livestock to graze plants they don't like will end in overgrazing of desirable species. Trampling, mowing or bushhogging is a better way to reduce invasive competition.

Utilizing multiple species of livestock and adaptive grazing management based on what you want and not so much against what you don't want, can alter the species mix in your pastures in a major desirable way within two to three years. This is especially true if soil fertility is adequate and rainfall is plentiful.

On badly degraded soils my first objective is to stimulate any and most every plant that will grow just to get the soil covered. But assuming we are working with good fertile soil, management is tailored to favor the plants I want more of. As a shepherd that usually means forbs and legumes. For cattlemen it probably means less weeds and more grass.

To favor native warm-season grasses we like to graze the cool-season grasses quite severely shortly before the warm-season plants begin to grow. The result is reduced competition for the native grasses.

Of course, if we didn't like native grass we would wait to graze those fields until the natives were growing and then we would allow the livestock to select the tender new growth warm-season grasses and leave the rank cool-season species to proliferate.

Grazing a pasture close to the ground and repeatedly returning after only a short rest period will generally stimulate white Dutch clover, dandelions and plantain.

Alfalfa thrives when grazed or mowed almost to the ground every time it blooms. This practice is hard on most grasses because they need some intact leaf area to fuel regrowth and/or a longer rest period to recover. If you wish to increase the grass component of the sward just leave a little more residual and the grass will respond accordingly.

High stock density in each paddock will result in non-selective eating and trampling. The goal with non-selective impact should be to have every plant either bitten or trampled, not every plant grazed to one height, which would end at ground level. Pruning the sward non-selectively gives all species an equal chance to regrow.

I like red clover, maintaining a robust stand requires grazing winter stockpile pretty short to stimulate germination followed by periodic non-selective grazing throughout the growing season. Red clover is a biennial, so a stand will disappear after two years unless new seed is produced and conditions favor germination.

These are a few tools we can use to generate a diversity of quality forages.

* Non-selective grazing when possible.
* Appropriate and varying rest periods.
* Leveling the sward once or twice a year.
* Timing grazing events to favor desirable species.

Don't forget, most plant communities are in a state of perpetual transition. The forage species you are growing today probably won't appear in the same ratios in two to three years unless your overall management suspends plant succession in its current state.

Livestock have the ability to renovate or deteriorate our pastures. Let's use the hooves and mouths of our ruminants to heal our soils and propagate desirable plant communities.

Chapter 28
A Garden in the Pasture

Did you know a healthy pasture is often home to many common fruits and vegetables? Here are a few samples from my pasture: asparagus, onions, garlic, strawberries, ground cherries. All these species have been domesticated and yet their wild cousins continue to thrive in our pasture ecosystems.

Of course, the list of wild edibles is not limited to species that have been domesticated. Many herbs and alternative salad greens are passed up as "weeds" and we are the losers for the oversight.

If you step into the edge of the woods the food list grows even longer. Silvopasture and woods edges are home to many fruits, berries, nuts, herbs and mushrooms. But I digress, I set out to tell you about my garden in the pasture but I got sidetracked just thinking about wild edibles.

My family started growing wholesale tomatoes when I was only a little fellow, perhaps seven years of age. I found this narrow discipline repulsive, but when I married Christina we continued the tradition of growing wholesale produce for several years, however we did make one change. We dropped the tomatoes entirely and diversified into several different crops. As our livestock business grew the vegetable business was forced to shrink and finally we phased it out altogether.

This led to a new dilemma, where was the garden? We were accustomed to selecting our daily vegetables from our diverse array of wholesale produce. Lamb without veggies is not my idea of a balanced diet. Alas we were spoiled by vegetables straight from the garden as opposed to a purchased product that has been wilting on a display table.

We didn't have time to grow a traditional garden, but images of trouble free produce continued to haunt me. I remembered how well the crops always performed the first year after a field was plowed with very little weed or bug pressure. In fact, the first season after breaking a healthy pasture sod, good yields could be expected without any fertilizer.

About this time I began to learn about soil aggregation and its positive benefits on plant health. Tillage destroys soil aggregates, reducing water infiltration and oxygen penetration. Among other things this increases dependency on fertilizer and irrigation. The result is a shallow rooted vegetable plant with compromised nutrient integrity, and a sharp increase in bug and disease pressure. The nutritional integrity of my peppers, radishes and onions matters to me because we are what we eat!

One day I moved a flock of ewes onto an old worn out produce field that we were trying to convert back to pasture. Standing there observing the sheep I realized they were having the time of their lives eating all those pesky weeds. Before long all that was left of the weeds was their naked stems. The sheep had obliterated the weed canopy exposing all the little grass plants underneath to the sun. In trade for the weedy repast the sheep deposited a heavy dose of natural fertilizer. Then a thought hit me. Here I am trying to find time to pull weeds in our little garden that is always in the same place, and the sheep are always looking for weeds to eat in our grassy pastures. How foolish I have been. Why not plant the garden a different place each year?

Assuming we would grow a huge quarter acre garden and plant it a different place each year it would take 80 years to get across the 20 acres of pasture closest to our house! It only takes an old garden site five years to heal over, again sporting a dense fescue sod. That means we could cycle back to the same location every six years and still reap all the benefits of structured/ aggregated soil and natural fertility from our sheep and cows.

Going beyond theory and putting my wild idea into practice, I set out to kill the rankest fescue on the farm. Since we wanted a chemical, cide-free garden the most difficult task would be killing the grass.

First we purchased a 15 ft by 300 ft roll of ground cover (plastic weed barrier). Next I cut it into three pieces 100 ft long and laid them side by side to cover an area 45 ft by 100 ft. To hold the ground cover in place we rolled out a bale of moldy hay on top of the ground cover.

We usually do this the first of April just as the grass is greening up. Thirty-five days later when the chance of frost is past we pull back one of the ground cover tarps and set out our first planting of melons, cukes, peppers, zucchini and yes even a few tomatoes.

The first year we did this I just knew the grass would take half the summer to die. We were pleasantly surprised after just 35 days to find the grass quite dead.

Planting directly into the soil without any tillage reduces weed germination. The hay that is used to hold the tarp down is scattered out on the garden as mulch to help retain soil moisture. Care to guess how many weeds usually germinate the first year? Not very many. In fact, most years we spend a total of two hours pulling weeds.

Even without fertilizer yields are great. In this model my crimson sweet watermelons frequently tip the scales at 25 to 30 pounds, and in 2023 we set a record by growing a 41-pound crimson!

This model has been a real breakthrough for our small scale commercial ground cherry enterprise that I refuse to quit. The first year we moved the ground cherries to the pasture garden they produced 14 pints of fruit each, and not a single plant died from bacterial wild. Historically we expected a 30% to 50% mortality rate for ground cherries and that was with regular spraying. In our garden in the pasture we haven't had to spray at all.

I like to locate the garden near one of the livestock watering points if possible. Pressurized water close at hand makes irrigation a breeze. We have found this production model has reduced irrigation needs significantly.

Positioning the garden along a fence makes it easy to fence out for the year. A roll or two of electric netting makes a quick effective multi-species barrier.

If the garden happens to lay beside a woven wire fence, then we plant cucumbers along the fence and let them self-trellis. If the garden lays beside an energized high tensile fence, then we plant the cucumbers by the corn and they trellis themselves on the corn.

Many different vegetables form mutually beneficial relationships, anything from providing a natural trellis to creating a microclimate for species that like partial shade. Remember the Indians' Three Sisters — corn, squash and beans?

Guess what happens in a garden plot the second year? It grows weeds, lots of weeds. The fungi/bacteria ratio has changed to favor early succession plants. Nature's cover crop is busy trying to quickly cover the soil. Instead of mulching the weeds, we simply move the garden to a new area and work the weed patch into the sheep's regular paddock rotation.

Moving our garden to the pasture can be summed up as a win-win-win. Below are parts of the sum:

* Reduced labor and fertility costs.

* Higher quality veggies from insect and disease resistant plants.

* Happy sheep every time they graze a weedy garden plot.

May your carrots grow long and slender, your radishes sweet and fat. When your sheep graze a weedy garden plot they will remember that!

Chapter 29
Drought Happens — Have a Plan

Let's face it, drought happens, but we can take steps to minimize the effect it will have on our pastures, animals and bottom line. The time to prepare for drought is before the rains quit. Have a plan and be ready for action.

With a well planned and managed pasture rotation we can measure our feed bank out in front of the flocks or herds. During dry spells and especially in drought, the recovery period for every paddock needs to lengthen. Assuming our normal, midsummer recovery period is 60 days, during drought I would like to lengthen that to 90 or even 120 days.

Forages grow slower during drought and it is critical to give the sward a chance to fully recover before regrazing.

Drought is not a time to graze a pasture down short. Overgrazing a paddock will result in a slower recovery. I know just how tempting it is to de-thatch a paddock, but even if there is no regrowth you can always rotate the flock through each paddock again, grazing the residual later in the drought.

Leaving significant foliage in each paddock will hasten recovery. A thick mat of vegetation covering the soil keeps the ground from heating up and evaporating precious water. Also, a big long wide leaf catches and absorbs greater quantities of dew, much of which moves down the water channels, oozes out through the roots into the soil and is taken back into the plant in the normal manner, traveling back up to the leaves.

It is amazing how heavy the morning dew can be during dry weather. And equally impressive is the way plant growth can be sustained. A study in Arizona showed that plants are able to obtain from dew the equivalent of 15 inches of rain in a growing season. Wow! Better leave plenty of leaves to capture the dew.

Dry weather optimizes animal performance so long as the forage stays fairly green. Also on the bright side, 10 inches of standing forage contains far more animal unit days during dry weather than 10 inches of forage in wet weather. This difference in animal unit days helps to extend the rest period.

As soon as I realize we are in a drought we destock by 15%. Early reduction of stock numbers means we get higher prices for our culls and have more feed left for the remaining animals.

By selling a few animals early on we minimize the risk of needing to sell a lot of animals later. Take the time to read that last sentence again and let it sink deep into your being.

Create a flexible stocking rate policy. Owning a flock or herd of expensive breeding animals that you cannot or will not dispose of is a good way to go broke buying shipped in feed. Every operation should maintain some animals in the good years that are easy to part with in the dry years.

Destocking should begin with your cull candidates, mediocre animals and anything that is not intended for reproduction such as wether lambs. You need to know how, what and when you will destock in advance of the actual crunch. Having a written drought destocking plan can take some of the emotion out of the process.

Why all this talk about destocking? Why not just buy hay and feed our way through? Because all the neighbors are experiencing drought as well, local hay is hard to find and too expensive to feed to mediocre animals. Trucking hay in from other areas is equally expensive.

Once you have made up your mind to destock, do it. Don't second guess your decision. If the rains come sooner than expected you may end up with surplus pasture, but then you won't need as much hay to get through the winter and that is positive.

Old timers here in Missouri tell about the 1930s when

it was so dry they were reduced to cutting trees to keep their cows alive. Every day the cows would meet them at the gate. The farmer would shoulder his axe, go to the woods and cut one tree down for every cow. Talk about drought! I have never seen it that dry, but again those old timers didn't have electric fence and the way I get it pasture management was basically unheard of.

We can't make enough money in the good years to balance out the money lost during the dry years if we try to feed our way through an extended drought with our normal stocking rate. And nobody knows just how long a drought will persist.

If you do buy hay to sustain part of your flock, pen them in one paddock or a dry lot to feed that hay. Opening all the gates, letting the flock wander across the dormant pastures nibbling everything down to the dust while eating hay will ruin your pastures.

When the rains do come, a severely overgrazed pasture will not respond to the moisture for one to two years. So, now you will be reduced to feeding hay through the wet years as well.

When drought happens, don't waste your time running in circles like a dog chasing its tail. Fetching water, feeding sheep and crying in your coffee won't ensure survival. This is a time when we need to make critical decisions on selling stock. Manage pastures so that the landscape doesn't deteriorate and the bank account is not depleted.

Chapter 30
Pasture First — Hay Second

I never cease to be amazed at how many stockmen place more emphasis on hay production than pasture management.

Livestock in general and sheep in particular prefer green living pasture over dry hay, and yet hay is what we prefer to feed. Hay loses significant feed value during the curing process. Conditions must be perfect to make excellent hay, not so with grazed pasture. Pasture quality can be high in spite of unfavorable weather. Livestock can harvest your forage, store the energy, and spread the manure all in one pass. How's that for efficiency?

For the record, I have nothing against making hay, but pasture must come first. Pasture should be priority and then if there is extra forage during periods of peak growth, it is okay to make some paddocks into hay.

Well managed pasture will usually generate far more animal unit days/acre than hay ground in spite of all the "wasted trampled grass." Like any "natural whole" the reasons for this production difference are many but there is one that stands out. A forage growth curve following severe pruning of a grass plant follows a more or less consistent pattern.

When the plant puts out its first leaf, solar collection is minimal. Plant growth gains a little momentum when the second leaf is added. Once the third leaf is developed, growth goes into high-gear. Rapid growth continues through the development of leaf number four. But, growth begins to slow with number five. As leaf number six develops, leaf numbers one and two begin to die and are drawing energy from the plant instead of generating it.

Peak growth occurs between the second and four-and-a-half leaf stage. Tailoring grazing management to keep forage in that rapid growth phase most of the time will give your pasture a big advantage over hay ground. Forage for hay is often let to grow until it has five or more leaves and is growing at a reduced rate. Then to maximize tonnage, folks cut their hay below the second leaf. Once again growth will be slow until the plant rebuilds an optimal solar panel.

As a side note, I have found that cutting hay a little higher and before it reaches the five- to six-leaf stage speeds plant recovery significantly. And yes, we graze the regrowth rather than make a second cutting the same year.

Soil life suffers when life above the soil is removed or reduced below optimum. The microbiome of the ruminant and of healthy soil is one and the same. It is shared from the soil to the sheep and from the sheep back to the soil. Microbes are shed on the soil from manure, in slobbers, and off the skin. Microbes are picked back up when grazing and lounging. The pulling, yanking action of grazing, stimulates soil life, whereas mechanical clipping has very little negative or positive effect.

Grazing "hay ground" in the spring/early summer is strictly a positive action. Grazing activity stimulates the forage and makes the plants tiller out instead of shooting up. The result is leafier hay. If grazing is timed right, seed production is significantly reduced, resulting in better hay.

One time we hired a neighbor to mow some first cutting hay for us. I was busy managing pastures and he had the equipment and the time. When he pulled into the field I wanted mowed, he laughed and said it wasn't worth cutting.

"Why?" I asked.

"Because it's too short! You should have seen my field I just mowed," he said.

We went ahead and mowed anyway. Two days later we square baled our hay; the neighbor square baled his hay too. And guess what, both fields yielded 60 bales/acre. But, we had grazed our field twice that spring and my hay was nearly as good as the second cutting. Our hay was short-stemmed, fine and leafy. The neighbor's hay was course, stemmy and nearly half seedheads. This same neighbor had fed hay for 30 days longer than us during the spring, all because he didn't want to graze his hay ground for fear of reducing hay production.

The real tragedy is that in spite of all the testimony from livestock performance, there are still so many stockmen who continue to feed hay late into the growing season. And for what purpose? To grow more hay!

The easiest way to reduce spring hay feeding is to put all your acres in the grazing pool. Yes, that includes all your "hay ground." Obviously if it is really wet, you wouldn't want to put cows on a newly seeded field or an alfalfa monoculture. Even sheep will tear up a fragile soil that has a lack of sod structure. Yet, we always get some dry weather sooner or later. The key to not tearing up pastures is management. And yet, a little hoof action will only stimulate a few weeds to grow and weeds are invaluable to sheep production.

Ideally, during early spring, we only graze the top third of the grass plants. This won't slow plant growth very much and will make the grass tiller out resulting in better hay or pasture later on.

If we began grazing our pastures when they reached the three leaf stage and make one full rotation, but on returning to the first paddock we find it has not recovered, then the pasture is overstocked. Better sell some animals or rent some more pasture. A recovered grass plant will have three or more leaves per stem that grew since the last grazing event.

The one time hay is more important than the pasture is when forage quality declines below hay quality during dormancy. My sheep, goats and cows want to graze "some" 365 days a year, but ice storms and deep snow sometimes prevent that option. Also in late winter, most of the nutrients in stockpiled forage have leached out, compliments of our wet winters. This deteriorated forage is not growth food, barely even maintenance food. At this point, offering some good green hay can really boost performance. If we wish to keep grazing stockpile non-selectively, offering legumes at 25% of the ration or good grass hay at 50% of the diet would be wise. Unfortunately, the effects of malnutrition don't manifest immediately. As a result, a lot of summer's problems are brought on by winter's deficiencies.

When we do feed hay, I prefer to feed it on the ground rather than in a bale ring. Square bales can be scattered out and big round bales just rolled out. Feeding on the ground reduces competition and helps spread the manure. There may be a little more waste feeding on the ground, but I also find that the ewes are good at sorting out the best forage, and most of the waste is stems.

As spring advances, at some point the sheep will "go green" and quit eating hay of any kind. That's fine, let them go for it. The hay you make later in the summer will be better quality as a result. Every time the sheep go green on me, I think of the story another shepherd shared with me once.

"A shipment of sheep from Scotland to Australia arrived off-shore in the land down under and before they could dock a heavy fog settled over the harbor. For two days the fog was so dense that the ship couldn't dock. All the sheep mysteriously quit eating hay, of course the folks taking care of them became quite worried. Then the fog lifted and the ship was able to dock and unload the sheep. They immediately started to graze. Mystery solved, the sheep-master realized the ewes had smelled the green grass across the water and quit eating hay, due to their extreme desire for fresh green pasture!"

Chapter 31
Orchards and Sheep are Symbiotic

The potential symbiosis between sheep and orchards as well as shepherds and orchardists is simply too great to omit. The relationship is especially beneficial if the orchard is being managed without "cides."

Wide tree spacing, allowing sunlight and airflow, is a fundamental step towards eliminating dependence on "cides." However, sunlight stimulates dense, understory vegetation that must be de-thatched before winter, enter the grazing horde.

While there are exceptions to every rule, sheep are the best species for this job. I'm not biased. I"m just speaking from experience. Cattle and horses can reach too much fruit and goats are a poor choice thanks to their compulsive browsing habits and ability to climb up in the trees.

The savannah-like setting of the orchard is a sheep's paradise. Indeed, the relationship is symbiotic. The trees provide shade and offer dietary diversity. In turn, the sheep control vegetation, prune the lower branches, clean up rotting fruit and fertilize the trees, inoculating the area with a fresh army of beneficial microbes.

As a shepherd/amateur orchardist, we time grazing events to encourage a desirable result for all involved. The whole thing comes full circle when we harvest all those bushels of delicious and rare varieties of tree ripened apples throughout August, September and October.

Many of the diseases that afflict orchards are said to reside in rotting fruit, something sheep love to eat, and fallen leaves that are slow to decay. A healthy population of soil life, a side benefit from the sheep's activity in the orchard, stimulates rapid decay of fallen leaves, theoretically reducing disease.

Even though this book is dedicated to shepherds, sharing a few orchard lessons we have learned would seem appropriate. After all, there are a lot of orchardists and shepherds who would benefit from a symbiotic partnership. And, if a shepherd wishes to eat an apple a day to keep the doctor away, your eating experience would be enhanced by tree ripened apples from your own orchard.

1. Plan ahead. Don't plant any trees until your soil is biologically healthy and active.

2. Select a site that will facilitate cold air drainage.

3. Windbreaks are desirable, especially when they are composed of trees, sheltering predatory birds.

4. Select disease resistant varieties with a root stock that is known to thrive in your soil type.

5. Limit sheep access to young trees, especially during the spring when sheep are shedding their coat. They will rub on the trees and may break them off.

6. Risk of sheep eating the tree's bark increases dramatically in late winter, a good time to remove the sheep. Sheep will leave the bark alone provided dietary diversity has been sufficient.

7. Irrigation during dry spells can make a huge difference in tree performance. The extra forage grown in the orchard during drought is always appreciated by the flock. Allow the soil to dry out between weekly waterings.

8. Tree guards need to be monitored seasonally. If they are loose fitting the guard may be providing a safe home for voles in the winter. If they are tight fitting they may be providing a damp habitat for borers during the summer.

9. Controlling vegetation with sheep adds to the orchard ecology. Mechanical de-vegetation, though sometimes necessary, is expensive and doesn't stimulate soil biology.

10. In an effort to promote fungal soil life, we stockpiled a 10-inch stand of grass in our orchard over winter.

Meanwhile the rest of our home farm was grazed off leaving three to four inches of residue. Consequently all the voles that lived in the 40 acres surrounding the orchard moved to the orchard for shelter and food.

Sheltered from ariel predators, beneath 10 inches of stockpiled grass the voles built runways from tree to tree and burrowed in around the root systems, eating the roots of trees that had been established for three to eight years.

When spring came and many of the vole damaged tress began to die, the full cost of that 10-inch mulch began to manifest.

Moral:

A. Grazing the orchard thoroughly during late fall exposes rodents to aerial predators, foxes and coyotes.

B. When entering the dormant season graze the orchard before the surrounding pastures. That way the voles will seek shelter and move out of the orchard.

C. Vole damage is minimal, perhaps nonexistent during the growing season. Vegetation under the trees doesn't need to be kept short all summer.

Considering the rising cost of land and a growing market for local, chemical free food, why not produce lambs and apples on the same acres? After all, the relationship would be symbiotic, reducing dependency on fossil fuels and "cides" for the orchardist and shepherd alike.

Chapter 32
Disease Prevention is Better Than Cure

Browsing through modern books on diseases of sheep would leave wannabe shepherds questioning the sanity of raising sheep. An in depth comprehensive study of viruses, bacteria and parasites can hypnotize the bravest soul.

But, nature is full of checks and balances. Why not have a big round of applause for the balance that viruses and parasites bring to the ecosystem? When a landscape or pasture becomes overstocked with a certain species, malnutrition opens the door for all the bad guys.

The diseases that re-balance the stocking rate will have names, but at the end of the day the result is the same: dead animals. Since there were animals present that were not able to meet their needs in their environment you could say nature recycled the weak, leaving resources for the strong.

Undoubtedly to the disappointment of many this chapter does not include a long list of ailments and cures pertaining to the ovine species. There are plenty of books about ailments of sheep written by qualified professionals. And there are certainly a lot of cures, some more effective than others. I don't want to get bogged down with the details of individual trials and treatments. Such a discussion would be impractical. Prevention is better.

If you followed my theme throughout this book you probably noticed a common refrain: If you control the breeding season to fit your forage context, adapted sheep can meet their needs on a biologically diverse landscape. As sheep numbers rise performance falls. As sheep numbers cross optimum sheep carrying capacity for your landscape, natural forces will seek to re-balance the ecology.

Did that sound familiar? If you are beginning to think, "Abram is narrow minded, belaboring his point on the importance of forage diversity," I would like to remind you that nearly 200 species of herbage is not narrow. In other words, diversity is not a silver bullet. It's hundreds of silver bullets.

I am not criminalizing confinement operations that choose to feed a narrow, restricted diet. I will suggest that there is plenty of unnecessary confinement production. Some systems certainly lend themselves to disease propagation more than others. Mud lot versus deep bedding or corn/soy rations versus alfalfa hay would be cases in point. I will also confess here that we operated a small confinement finishing enterprise for several years. Thin lambs could be bought in September and then be resold in January.

By playing the annual price cycle the lambs would bring more per pound when sold than when we bought. Plus, we got paid for all the pounds they gained during the 90 day finishing phase. The manure and bedding generated from this enterprise was spread on our pastures to grow more forage.

Indeed, I feel no guilt about my CAFO involvement of the past, nor do I fault anyone else for doing the same. If we hadn't purchased the lambs they probably would have ended up in a less friendly environment. Regardless, they would have been fattened on corn.

We discarded this enterprise after a few years, mainly because local competition for feeder lambs became so strong that there was no forward margin left in the market. Also, a spike in grain and hay prices diminished the profits from weight gain during the feeding period.

Let's get back to our discussion on preventing disease.

All sheep are NOT created equal. Sheep have been selected and developed for a wide range of environments and purposes. Some sheep fit my farm, others don't. Those that don't fit fail to thrive. I cull the misfits and move on with life. If the majority of your sheep fail to thrive it would be wise to take good look at your entire management strategy. If management is not the problem consider the stocking rate. And if that isn't the problem, then you need a different breed, a radically different breed that is adaptable to your context.

Every rapid rise in disease is our fault. That being said, here are some potential problem makers.

Stress. You don't function well if you are stressed. Decisions may be rash or inappropriate (sheep have a lot of decisions to make when selecting their diet). If your worry and frustration continues you lose sleep, lose weight and eventually get sick.

Rapid changes may cause stress — new location, forced change of diet, exposure to new sheep, change of management, etc. Easing changes is helpful when moving sheep to a new area, allow an adjustment period. Pressing sheep into an environment to which it is not adaptable causes stress with no hope of relief. Phenotypical extremes may cause stress, e.g. tall, narrow or long bodied.

Toxins. Any source of undigested mineral can be a source of toxicity. Never force or entice sheep to consume salt or mineral. If they won't eat it, they don't need it. Iron, responsible for the pretty pink color in some natural salts and many mineral blends can wreak havoc in an otherwise healthy flock. Water sources are often high in Iron but this element tends to settle, so drawing water from the surface of ponds or reservoirs should yield a lower Iron content. We avoid any pink salt or mineral and any water source that appears orange, like rust. Again, draw water from the surface. That is where animals drink under natural conditions.

Chlorine used for bacterial control in water sources is a toxin. Of course bacterial contamination is a problem too, but sunlight tends to kill harmful bacteria.

If water quality is sub-par, offer shade, thereby keeping the sheep cool to reduce their water consumption.

Antibiotics are a toxin. After all the word could be translated as "against life." Antibiotics kill life a little at a time. Rare strategic usage may give individuals a second chance, but ongoing usage leads to a downward spiral. The already under-nourished body now has the double responsibility of detoxifying this foreign substance while trying to metabolize enough forage to get on top of things, all while trying to rebuild the microbiome.

Toxins may come in many forms. Here is my thesis for protecting my flock. If the product was not available for sheep in the year 1500 they don't need it today and may be better off without it.

Lack of carotene. Carotene is the basis of vitamin A. Sheep that have healthy levels of vitamin A acquired naturally generally exhibit strong resistance to infections of any sort. Carotene content declines in dormant forage due to oxidation. The same thing happens with hay and silage. As a rule green forage and green hay are adequate sources of carotene. A sheep may be able to store enough vitamin A to sustain her for up to four months. Feeding only brown forage for longer periods of time may cause problems.

As long as forages grow and rivers flow we can raise herbivores, but the herbage that grows and the water that flows will ultimately determine which species of herbivores will thrive. It is well documented that cows thrive on grass, the likes of which a sheep will only survive on. In contrast, sheep thrive on a diversity of forbs, grasses and woody plants. This concept is fundamental to disease prevention.

In conclusion, a happy animal sustained by its natural diet does not stress easily and is more tolerant of change, short periods of malnutrition and toxin exposure.

If you find your flock in the lurch, the steps to recovery are the same steps taken for prevention, perhaps backed by extra TLC and strategic, expensive treatment.

Chapter 33
Parasites — Whose Fault?

Food is everything. All life is either eating or being eaten. I view the process as a magnificent, perpetual struggle of life, balancing and counter-balancing stocking rates of bugs, worms, herbivores, carnivores and omnivores, bacteria, fungi, reptiles, weeds, grasses and trees. The list goes on.

A reduced food supply equals a reduced carrying capacity equals a reduced stocking rate eventually and reduced performance immediately.

In my early teens my brothers and I spent countless hours trapping, hunting and shooting Norwegian rats and English sparrows. Rats in particular can be quite destructive. After several years of ceaseless war on these pests I realized my only meaningful accomplishment was marksmanship. I had become a crack shot. All that carnage failed to reduce the population simply because the carrying capacity for rats and sparrows was still the same. Grain is a major component of these pests' diet and they had unlimited access. Protecting all the feed sources would have nearly annihilated the pest population in a matter of days. Indeed, food is the basis of all life.

All life is interdependent, a part of the food chain. No species is exempt because we all impact our environment. When the stocking rate of any one species exceeds the landscape's carrying capacity, for that species' natural functions are going to try to rebalance the population. Since sheep are generally confined to your farm and can't just walk down the road to a neighbor's weedy pasture, immunity and resilience of the flock is going to suffer when stocking rate exceeds carrying capacity.

The plant life in your pasture ultimately determines which species of herbivoes will thrive.

Stress and malnutrition are the passwords for internal and external parasites to atack a sheep with lethal force.

Every environment, in fact every farm is unique in some way. The sheep that are adapted to one environment may not be adaptable to another environment. A lack of adaption predisposes a sheep to stress and potentially malnutrition. Given a chance, parasites will recycle the unfortunate ewe or flock unless some other predator takes her out first. As sheep numbers fall, forage selection for the remaining flock increases. With this increase of nutrition comes a rise in parasite resistance among the survivors, provided nutrition was the limiting factor.

I have come to believe parasite resistance of itself is not genetic. If it is genetic, why would resistant sheep succumb to parasites following a change of environmental pressures? But, genetics do determine one's adaptability to an environment, so we gain parasite resistance by default when we select sheep adapted to our environment.

Here at Still Waters, selecting and propagating sheep adapted to our environment has resulted in a flock that is parasite resistant within our context. This is simply because they are happy, not stressed, and are able to meet their nutritional needs by grazing the forages growing in our pastures.

Keeping the flock worm free is not feasible. Sheep can handle a moderate parasite load. Sterilizing the gut to the point of worm eradication would likely cause digestive failure, thus increasing susceptibility. A few worms won't kill a sheep. If every animal died that had been infected, parasites would lose their host species and die out; however, heavy infestations of internal parasites are very lethal to susceptible animals.

The greatest risk from stomach worms, also called round worms, is during or shortly after a period of hot muggy weather. Anemia is usually the first symptom. Anemia shows up as a pale grayish color of the inner eyelids and gums, which in healthy animals are a healthy pink color.

Pink nosed sheep are especially easy to spot. Their

whole nose will appear ashen, deathly pale. This disorder is the direct result of internal parasites drinking the sheep's blood, up to a pint a day. Yes, just draining their life away. Red blood cells may be reduced from acute haemonchosis, to a fatally low level in just a few weeks. Post mortem examination reveals a lot of chocolate colored fluid in the stomach, which is blood altered by chemical action. Flock-wide anemia calls for immediate action, no time here to rebuild parasite resistance. Deworm them and then work toward solving the issue.

Aside from parasite resistance, there are two basic ways shepherds can reduce parasitism without regular deworming.

1. Deterrents. Various products such as copper sulfate can be fed to the sheep in conjunction with salt. Basically the sheep are consuming a low grade dewormer every day; however, anything that is toxic to worms will also reduce digestive function. Herbal compositions fed in the same manner are probably a safer option and are becoming a popular tool.

Weeds, which are actually herbs remain at the top of the list for parasite deterrents. Many weeds are anti-parasitic and also build the immune system.

2. Reduced exposure. A rest period for each paddock of 50 to 60 days will reduce the presence of infective stage parasites. Parasite larvae may become infective within seven to 20 days depending on environmental conditions. Warmth and moisture hasten development while drought and freezing temperatures retard development.

While many parasites may enter their host orally, some species can enter the body through the skin. If you are serious about reducing exposure, sheep should not be allowed to use the same bed ground for more than five days in warm weather.

Some producers claim to reap huge benefits from breaking the parasite cycle with an alleged dead-end host such as cattle. I believe the greatest advantage cattle bring to the table is their propensity to germinate weeds and their nonselective preference for grass, the plant every sheep pasture is long on. Indeed, continual daily exposure to parasites in a set stocked pasture is not a big threat if that pasture happens to contain a wide variety of plant life.

Parasites are just another opportunistic predator, a part of natural balance and in some cases a species that must be killed in order for sheep to have a second chance. I like to think of this predator as a template for managing pastures, stocking rates, and selecting adapted livestock.

Folks often huddle behind the notion their area is specially predisposed to parasite issues. That refrain is heard East to West, North to South and at all elevations.

If stress and malnutrition predisposes sheep to parasites in every locale, then the opposite is also true!

Learn to mitigate stress in your flock and remember, it's eat or be eaten. Food and water is the basis of all life, so let's feed our sheep in a way that they can largely resist feeding the worms.

Chapter 34
Multi-Species Benefits

From bugs to plants to herbivores, the more diversified the community is the greater that community's independence. Said another way, single species or individuals are incomplete and consequently more susceptible to outside pressures.

In trying to simplify our lives with monocultures and single species livestock production, man has complicated his existence with pharmaceuticals, herbicides, excessive equipment and all sorts of expensive supplements and substitutes. That could be replaced with simple, affordable forage diversity.

The best thing about multi-species of grazers is the ability to match the grazing species to the existing forage species. When sheep are happy grazing the forages allocated to them, they become parasite resistant. When cows are happy grazing the forages suited to their metabolism, disease pressure falls and reproduction rises. Health and reproduction are indicative of adaption. Adapted livestock are low maintenance, enjoyable and profitable. This holds true for all species. The catch here is no species of grazer is adapted to thrive on every herb of the field, in whatever quantity or ration that herb may appear.

Record high sheep prices and low commodity cattle prices are encouraging sheep producers to ramp up production. That is fine as long as your forage base is suited to the metabolic needs and capabilities of your sheep. On the other hand, cattlemen are often afraid of adding sheep or goats from fear of increasing their workload. My question for those folks is how many hours and dollars do you spend controlling weeds and brush? I personally know cattle ranchers here in Missouri who maintain a flock of sheep and do nothing with or to them except to round up lambs and sell them once a year. The cattle are their centerpiece but the sheep are quite profitable and do not add to their workload at all.

Here at Still Waters sheep are the centerpiece and I aspire to run as many ewes on a per acre basis as my forages will support. As a result of my aspiration I move the sheep daily most of the year. As long as I keep sheep numbers in balance with sheep friendly forages no other labor is required. So how does my labor budget compare to a cow-calf enterprise with comparable pasture management?

Sheep in my paradigm require twice the labor of cow-calf. Sheep need two strands of wire to gain optimum forage utilization. Cows need only one wire. The extra labor required for sheep is compensated in the price difference. This past winter I sold 60-lb fat lambs for $4.00 lb on the open market. Fat cattle at the time were bringing $1.40 lb.

My focus on sheep actually increases my interest in cattle. I have found that taking the cattle out of the environment has led to a decline in sheep carrying capacity and sheep performance. To gain benefits of cattle impact I custom graze dry cows May through August. By sending the cattle home in August we are matching our stocking rate to the forage growth curve. We also own a small herd of cows that stay year around. I use them to graze over-ripe stockpile and germinate weeds for the next growing season.

Remember the key to multi-species success is matching the grazing species to the forage species. The accompanying chart gives some basic guidelines on grazers' forage preferences. An inventory of your forages will tell you which species of grazer should do well on your farm. Say you want to raise sheep, the first question to answer is do I have access to sheep-friendly forages? This concept applies to all livestock.

Personally, I enjoy goats but my forage options limit my goat herd to 10% of our sheep numbers. Cattle do well on the grass overlooked by the sheep. Without the heavy impact of cattle we would soon lose the forb component of our pastures.

Approximate forage preferences of herbivores

Animals	Grass	Forbs	Browse (wood)
Deer	10%	30%	60%
Goats	20%-45%	5%-15%	40%-75%
Sheep	50%-65%	20%-30%	15%-30%
Cattle	85%-90%	5%-10%	5%

One important consideration often overlooked is that all species of herbivores increase their grass consumption in the dormant seasons and decrease broadleaf consumption. As a result sheep, goats, cows, horses and pigs are competing more for the same resources. That being said my sheep eat a considerable quantity of red cedar in winter and my goats live on buckbrush and other browse once the grass is covered with snow.

Owning several species is of little value if they are managed as separate units and always remain on their own corner of the farm. Throwing all the stock into one big "flerd" isn't necessary and may well be impractical (pigs will eat lambs, cows will charge guard dogs) but ideally every species will rotate across the whole farm at some point in time. This is called organized chaos, it's great fun.

Adding more stock can seem daunting considering winter chores. After a couple of years trying to do it all, we gave up farrowing our own pigs so that we could go pigless in winter. If the stocking rate should be matched to the forage growth curve then the chore load should also be matched to day length.

One way to reap the benefits of multi-species is to team up with a youngster who lacks a landbase. This scenario can become a win-win-win. The youngster will be around when you need an extra hand or want to go on vacation. More people on the land is a plus.

For micro operations too much diversification can lead to financial disaster, especially if you are trying to mimic larger operations. The micro farm must adapt multi-species management to fit their own situation. Teaming up with another micro farm is one way to gain the benefits of more species. Great things can happen when farmers work together as a community.

In the end whether it's farmers, livestock or forages, diversity is multiplication. Allowing forages and livestock to diversify and multiply has actually simplified my life and increased my profitability.

Chapter 35
Goats — Less Work Than I Thought

While the word shepherd is in no way connected with goats, many shepherds, me included, keep a goat herd for ecological reasons.

In many ways goats are similar to sheep in form, size and production, with a similar life span and both species can be susceptible to the same diseases and parasites.

In spite of their similarities, there is one striking difference — goats eat far more woody plant material (browse) than their sheep counterparts. This fundamental need for brush is my main reason for owning goats.

While some of our pastures are much too clean, other pastures have unending brush encroachment. By adding goats to the species mix we are gaining broader and greater forage utilization on these brushy farms.

My original fears of goat trouble have proven to be ungrounded, once we got the context right for goat success. To make a long story short, our goats are easily contained with the same fencing system we use for the sheep, kid once a year in the spring on pasture, receive only salt and hay as supplements and are otherwise hands off. While the children have made pets out of their favorites, the only handling required is for culling and selling the kid crop.

Goats that are new to electric fence require a little more training time than sheep, but the kids born here on our farm self train shortly after birth.

Of course goats, like sheep, are easier to fence if there is plenty of suitable forage inside the fence. A rough and brushy pasture will serve as buffet, drugstore and gymnasium all in one. Being natural athletes the herd will spend hours playing on stumps, logs and steep banks. Pen a few goats in a small boring paddock and the chances of them getting sick or escaping their confines is amplified.

While cool, dry weather doesn't phase our goats, if it's raining they do appreciate some kind of shelter. During spring summer and fall, brush and trees are sufficient (cedars are their favorite). During the winter we take pity on them and allow access to a roof and three walls.

Ranging out from their winter shelter our goats will continue to nibble on dormant twigs, leaves and berries. Acorns are a delicacy and the goats grow fat as seals during a good acorn year.

Goats will increase grass consumption during the dormant season, making them more competitive with sheep and cattle for stockpiled forage.

Given enough area to browse, goats can just about live without hay. In our record winter, gross total hay consumption per goat was four pounds, however, if really high quality hay is available, consumption will be much higher. Winter nutrition does have a big impact on summer resilience and the size and quality of the annual kid crop.

I used to believe goats were inherently a lot of work — excessive parasite, hoof and maternal issues. I wasn't entirely wrong. That is the story in many goat herds. In spite of premonitions by selecting goats that are adaptable to our environment with the addition of winter shelter we are weaning upwards of a 200% kid crop annually, trimming no hooves and deworming no goats.

Breed isn't important to me — Spanish, Pigmy, Kiko — just so they are adaptable to our context and produce a healthy set of kids every year.

I was warned that goats are poor mothers. Experience has proven they can be, however I find they are much more patient and sneaky about caring for their newborn than are ewes. A goat has the patience and wisdom needed to feed three or four kids, making sure each one gets its fair share. Sometimes they will park two kids under a cedar tree and hide the third kid by a log. The kid/kids are then fed separately, which means there is a nipple for everyone.

Newborn kids may appear lethargic when compared with lambs. Goats will often hide their kids for days, much the way a deer will hide her fawn. Don't worry about it. The vigor of youth cannot be suppressed forever. The kids will be up playing and following the herd soon enough.

Since the goats are very selective browsers, targeting a plethora of generally inaccessible and unwanted forage, I don't rotate pastures as frequently as we do with sheep unless they are co-grazing. And we don't rotate at all during the 18-day kidding season. Parasites haven't been a problem in spite of staying in the same paddock for 10 to 20 days at a time. After all, many woody species are anti-parasitic, plus parasites are much more prevalent in the bottom of the sward than they are on the tips of the brush and weeds that goats thrive on.

Time has proven that adding goats to our species mix was a good decision — less work than I thought. With forage preferences and market cycles differing from their sheep counterparts, goats have added a unique complementary and profitable twist to our business, a business fueled by free sunshine and water, producing the world's best food and healing the land.

Are goats your number one niche? That will depend on whether you have access to brush. Deprived of their native diet, goats will likely be more work than you think.

Chapter 36
Reliable Guardians for the Flock

Let me shorten your learning cure. If you plan to raise sheep on pasture, protection from predators will be necessary sooner or later. Effective protection can be maintained in numerous ways, but three options stand out from the rest.

1. Predator exclusion fencing. This option can take several forms. Night time penning in a predator proof corral, micro flocks contained 24/7 in small moveable pens or "sheep tractors" and elaborate, predator proof perimeter fencing. Some shepherds find electric netting to be an effective exclusion fence, provided local canines are otherwise well fed.

Exclusion techniques work best in counties that are sparsely populated with sheep. I got along well as a boy using exclusion techniques, but that all changed when several new shepherds entered the sheep business nearby. The policy with some of these new shepherds was "my sheep are just small cows." Needless to say, local coyotes, foxes and stray dogs nearly annihilated their lamb crop and then proceeded to devour their mature ewes.

With new enlightenment on the nutritional values of lamb and mutton, local predators have become so bold that night time penning is no longer sufficient protection in our area. Coyotes and foxes are very adaptable. If a flock is penned at night they switch to feeding by day. This local change has reduced predator exclusion options to full fledged exclusion fencing.

2. Predator deterrence. Bells, radios and strobe lights have provided some shepherds temporary relief, until predators adjust to all the ruckus.

3. Guardian animals remain the best deterrent of all.
My first experience in this realm was with a llama, an effort
that didn't last long, so I won't waste a lot of ink on the subject.

Next we considered donkeys, but it soon became obvious that donkeys could only protect a very limited number
of sheep and that number was reduced even further by rough
brushy terrain.

My thought was that if we could use herbivores as
guardians they could subsist off the same landscape as our
sheep. However, both llamas and donkeys can't physically go
everywhere a sheep does. Plus, the fodder required to support
one donkey will easily feed two ewes and their lambs. In my
experience, the commercial value of those two ewes' offspring
will buy enough dog food to support two guard dogs year-
around.

Two guard dogs may effectively deter predators from a
flock of 500+ head, provided the sheep are contained in a one-
to two-day sized paddock.

I have no idea how many donkeys or llamas would be
needed to replace two good dogs. I am sure it would require
enough donkeys to make you say OUCH when you pay your
hay bill. I'm sure you can see where I'm going with this guardian discussion — guardian dogs are the answer.

Fortunately, we crossed the steepest part of our guardian dog learning curve prior to the local predators becoming
widely educated on the super food called mutton. I just wasn't
about to take a chance on losing sheep to coyotes.

My early concept of guard dogs was that any dog from
a known guardian breed could be exposed to sheep and trained
to be a reliable guardian. I won't bore you with the details of
that fallacy, but I will share some of the results.

I had fond childhood memories of playing with big,
furry, white dogs from a breed called Great Pyrenees. Among
other things Pyrenees are considered a guardian breed so I
decided to give them a try. After all, Great Pyrenees pups are
cute, common and inexpensive.

Unfortunately, my new guardian dogs had no interest in

sheep except occasionally to chase them for sport. If the dogs could get out of the sheep pasture they never returned on their own. The Pyrenees' loyalty to the flock was zero.

With the first lambing season my Pyrenees rapidly developed a taste for lamb. My new guardian dogs had become predators and would kill just for fun. The only thing worse than having no guardian is a bored dog penned with the sheep and preying on the flock.

Efforts were made at every stage to train these rascals to behave and to stay with the sheep, but alas, my efforts were in vain. How are you to train a dog that is unsupervised 23 hours each day? As a result every Pyrenees on the place soon found an alternative occupation. Was I too hasty? I think not. Those mutts were LGD in name only.

About that time I attended a pasture walk on a multi-species operation with some scale and noticed that the guardian dogs were in the sheep pasture with the sheep even though the pasture was big and the fences weren't dog proof. In other words, these dogs had the option of finding another home but preferred to stay near the flock.

I asked the owner what his secret was. His reply, "They are all at least half Maremma and we don't pet them."

After going home I couldn't get the image of those dogs out of my mind. The way they interacted with the flock was almost a species apart from the Pyrenees. One day I met the same shepherd at another pasture walk and asked him if he had any dogs for sale, he said, "Not presently, but I will have another litter soon. They will be ready for work when they are six months old."

The price for a six-month-old LGD was equivalent to three good ewes in the prime of life. I promptly reserved a male and later named him Pup. In due time Pup was introduced to our sheep. The flock was penned in electric netting for the occasion and we simply took the young dog to the flock and released him. Pup slunk over to the sheep and they ran away. He followed, passed through the middle of the flock and began to lead them.

Pup worked faithfully for eight years, on duty 24/7/365, well, almost 24/7. He would occasionally slip out of the sheep's paddock to feed on deer, pork and beef offal during butchering season, but he would always return to the flock on his own. In fact, the least disturbance in the flock would send him pummeling over the hill to check on his charges.

If the sheep were happy, Pup was happy and if they were disturbed, he was disturbed. The kinship he felt for his dependents was obvious, however, as much as he loved his sheep family, Pup was still a dog and occasionally likes to play doggy games.

If another dog was present in the flock, they would take out their extra energy on each other. If alone, the sheep would receive his unwanted actions. These actions culminated in several dead lambs when Pup was four years old.

I corrected the behavior with a twirl stick attached to a collar and made sure Pup always had a companion dog. The problem never recurred. Moral, dogs work best as pairs or trios.

The local predator issue has become so bad that sheep are not even safe in our backyard unless a guardian is present. In recent years we have tried drift lambing without giving the dogs access to all the paddocks. Foxes slipped in and killed several lambs in broad daylight.

The predator issue isn't limited to educated canines. When I was a boy we rarely saw a bald eagle. Today eagles are thick as thieves and yes, they will steal lambs. I have witnessed that firsthand. Fortunately, our guardian dogs have gotten wise to the eagle threat and seem to be effectively deterring them.

Mountain lions have been reintroduced to Missouri during the past 20 years and are proliferating. Black bears were recently released in Southern Missouri. Both of these marauders pose a growing threat to unguarded flocks.

Some of the very species that once carried a bounty on their head are now being reintroduced and are protected by law. One would think that if feeding the world is really a concern, we wouldn't share our food resources with unneeded predators.

With public sentiment favoring reintroduction and

protection of more and more predators, effective predator deterrence is and will remain a necessity to sustain a profit margin in the sheep business.

Several years ago recognizing the need for more reliable guardian dogs, we began raising LGDs. This has added an exciting element to our business and provided a lot of insight into guardian dog psychology. Success has been realized from a combination of dog genetics and shepherd management. So far, I have never had to repurpose a dog that was born and raised on our farm. We have never knowingly lost any sheep or lambs to predators when two or more dogs were bonded to the flock and the flock was contained in a reasonably small pasture. I can sleep well at night knowing our reliable dogs are on duty.

Best of all, I know my dogs enjoy their job because they have the option to roam and never return. However, that never happens because they are bonded to their flock. Now that is devotion.

Chapter 37
Guardian Dog Care and Breeding

Dogs are referred to as man's best friend, and for a relationship to be sustainable it must be symbiotic. As a result, man is dog's best friend. This bond is instinctive for both species. Over centuries shepherds have developed breeds of dogs that will bond to livestock or man.

Notice I said (or man). It is difficult to maintain an effective bond with two species when one lives on pasture and the other in a house. I suggest that if you want your guardian dog to maintain an effective bond with your sheep you will have to resist your own instinct to bond with your guardian dog. Of course, if you can't resist, you have the option to live out in the pasture with your flock.

Well bred and bonded dogs seem to view their flock as family. The family unit may include two or more species, such as sheep and goats provided they live in the same pasture.

A dog may play with or tease family members but won't cannibalize. Dogs won't eat dogs, and when truly bonded to sheep they won't eat sheep unless the meat is tainted and rotten. To prove my theory I offered fresh ground mutton and fresh ground pork to my dogs. They wolfed down the pork and never touched the mutton.

Notice the qualifier is the extent of the dog's bond to sheep. The instinct to bond is set down at the point of conception but the actual bond occurs between two and four months of age, circumstances permitting.

Allowing young pups access to dead sheep is not a good idea. However, there is nothing wrong with feeding raw meat to pups and adults LGDs. I feed deer, beef and pork offal to our dogs whenever possible.

If you have access to offal or trimmings from a butcher shop go for it. Butchering waste is cheap healthy dog food and will give your guardian dog's coat a shine, unequaled by a diet of kibble.

The bonding instinct is very much a genetic trait that varies in different breeds and families. Of the six guardian breeds I have worked with and observed, the Maremma has the strongest bonding instinct. The Great Pyrenees has the least desirable attributes, albeit they make good pets. And the Anatolian Shepherd is the most aggressive towards perceived threats.

A good dog should place himself between the flock and the perceived danger and in that position will run back and forth barking. I don't want a dog that is inclined to chase coyotes over the hill and into the next county. When a dog follows a decoy coyote more coyotes will come in from behind the flock.

I want a dog that will develop a close bond to the sheep, bark with all its might when aroused and only seeks to hold predators at bay. We have found this happy mix of traits by breeding a mixture of 3/4 to 7/8 Maremma and 1/4 to 1/8 Anatolian Shepherd. Purebred Maremmas work well too.

A common question with new shepherds is shelter for the dogs. Here is my suggestion. Build a portable dog house that you can move with the flock, if that makes you feel good, but don't be surprised if your dogs choose to bed down with the flock. If the weather gets really bad the dogs will snuggle up to several ewes and stay quite comfortable. Puppies however, are a different story. They need shelter.

Our dogs mate and whelp on pasture while serving as guardians. After a nine week gestation period, the female will select a sheltered place to whelp. Options may include brush piles, haystacks or hollow trees. One time a litter was born in a big hollow tree, five feet across at the base with only a 10 inch hole for access. Snug and safe to be sure.

Of course the sheep flock continued their paddock rotation, which soon distanced them from the new litter by a quarter mile. Sadie was torn between her interest in her puppies and loyalty to the flock. She began to divide her time between the two before the puppies even opened their eyes.

We decided to move the puppies to the flock and give them portable shelter, but alas, I couldn't get the pups out of the tree. Trying to escape the heat of mid-June the puppies had crawled into the furthest recesses of the tree. Our oldest daughter, age three at the time, and ever eager to handle puppies volunteered to crawl in through the 10 inch opening and pass the puppies out to me.

Pups don't show much interest in sheep until they are about eight weeks old, but exposing them to the flock from the day they are born will hasten the bonding process. The sounds and smells of the flock will be all the puppies ever know.

I like to place the puppies' shelter near the sheeps' shade or water sources to ensure at least some interaction daily.

The earliest interest in sheep is often stimulated by the actions of older dogs. Pups will follow the other dogs around copying their behavior. Even if no adult dogs are present, well bred puppies should bond to the sheep by four months of age.

There are two pup training theories that I believe have ruined a lot of otherwise good dogs.

1. Pups should be bonded to lambs first.

I suggest the opposite approach. Start them out with an adult flock. Training adolescents is a job for adults, not other adolescents. Both mature ewes and mature dogs will give young pups some much needed social lessons resulting in better dogs than their kennel-raised counterparts.

2. Pups should be introduced to the flock by degrees over time and always with human supervision.

Again, I would suggest a different approach. Here is why. The daily excitement of being walked on a leash through the pasture is too much excitement.

Puppies should be introduced to the flock by the time they are eight weeks old if not before. Keep them in a kennel in the middle of the pasture for two days. This will allow them to become familiar with the sights, sounds and smells around them. This introductory phase also allows the older dogs to get acquainted through the kennel barrier. After two or three days release the puppy/puppies and quietly leave the pasture. The ewes will train your puppies and you don't want to interfere with the bonding process.

Buying dogs is a lot like buying sheep. Buy from a flock breeder that manages the way you plan to. Take time to observe potential guardians before buying, unless you know the owner. One day some folks from town asked me if I had any bottle lambs for sale. As it turned out they had several dogs that they wanted to sell and they figured that by introducing a few bottle lambs into their yard their dogs could be marketed as "livestock guardians."

Well bonded dogs recognize and prefer to hang out with the sheep in their own flock, relocating a good dog to a different flock nearby, he will likely return to the original flock. After all, the first flock is considered family and the second flock is distant relatives. When a dog is relocated he/she needs a little time to accept the new flock.

When we sell a puppy or dog of any age these directions go with them. We follow the same procedure when buying new LGDs.

1. Place the dog in a cage or kennel in the middle of the pasture for 24 hours. Feed and water him/her and then leave it alone to absorb his surroundings.

2. Make sure the flock is contained in a pasture small enough that your sheep are always within close proximity of the dog.

3. After 24 hours open the cage and quietly leave the pasture without bothering the dog. It will exit soon enough.

We have never heard of our dogs leaving the county when these instructions are followed. Instead we get rave reviews on how the dogs head straight for the sheep and stay with the flock.

The time spent separated from sheep is quite stressful for a truly bonded LGD. When they arrive at their new home they are simply overwhelmed to find sheep waiting for them.

164

But an immediate release may result in the dog trying to find its way back to the original flock. Give them 24 hours to get oriented and they will do fine.

A dog that is accustomed to sheep can be rebonded to goats and vice versa but it takes a little time, 48 hours at a minimum.

Drawing from personal experience and from connections with many other shepherds, it has been my observation that LGDs work best as pairs or as a trio.

Good dogs that turn bad and puppies who develop vices are often just lonely, bored single dogs that to fill the void start picking on sheep. By keeping two or more dogs together they tend to play with each other rather than the sheep. Having a doggy companion does not reduce their bond to the flock at all.

If your lonely, single dog develops a bad habit, try correcting the behavior with a twirl stick. Here is how they are made. Take a straight stick approximately 1.5 inches in diameter. Or use an ultra light metal pipe. The stick should be between 18 and 30 inches long depending on the size of your dog. Attach a chain to the middle of the stick so that it balances when suspended and fasten the other end of the chain to your dog's collar. Adjust the chain so that the stick is suspended four inches above the ground when the dog is standing.

This apparatus provides immediate discipline for playful behavior, but allows the dog to express submissive, friendly behavior towards the flock.

A twirl stick will correct wayward dogs but will never turn a genetically inferior dog into a trustworthy guardian. We have used a twirl stick on three dogs who "turned bad." After they straightened out we gave them another dog to keep them company, and the problem never recurred.

I shudder when a new shepherd tells me they have an eight-month-old pup that they plan to introduce to the sheep.

If you want your dog to be a guardian when it's old, start the training process when it's young. Generally, puppies should be exposed to a flock by the time they are weaned, if not before. If the ewes knock it around, the pup will become a better dog for the respect learned.

How much a dog can be handled without spoiling it is a matter of debate. Honestly, I believe it depends on the dog. Some love attention. Others are wild.

Occasionally petting your dogs won't ruin them, but it certainly won't improve their devotion to the flock, and it's this devotion to their extended family or "pack" that makes a guardian dog valuable.

A well bonded dog does not require dog proof fencing to ensure it stays with the flock. Even the best dog may roam short distances from the sheep, but will always return to their flock on their own.

To allow our hard working dogs 24/7 access to kibble dog food when raw meat isn't available, I designed a portable feeding station with a creep gate to exclude the sheep. Sheep will develop a taste for kibble quite rapidly — an addiction that can't be healthy for the sheep, and it's expensive for the shepherd.

Here is the blueprint:

1. Cut a hog panel 140 inches long. This panel will be curled into a half moon measuring 45 inches across the face.

2. Using 2 X 4 lumber, build a gate 45 inches wide by the height of your hog panel, typically 34 inches.

3. Leave an 8.5 inch opening between the first and second horizontal boards. This is the dog's entrance.

4. If your dogs refuse to enter the feeding station, temporalily enlarge the entrance to 12 or 14 inches.

5. Place your dog's dish or self feeder in the very back of the feeding station. This discourages ewes from trying to sample the kibble.

When I got my first LGD we hand fed the dog daily. I would simply deposit the kibble in the pasture, whistle for the dog and walk away. He wouldn't eat while being watched. Upon adding several dogs the need for a feeding station became apparent.

The first time I built a portable feeding station we left the bottom board off and one morning I found three sheep trapped inside the pen. They had stuck their heads inside the dog's entrance, gave a heave forward and were trapped. By placing a board at ground level they were effectively excluded.

Since this design is lightweight, and I like it that way, we either fasten it to a tree with a dog collar or if the ground is soft, fasten it to a 14-inch rebar stake that can be pushed into the ground.

For our operation, Livestock Guardian Dogs are indispensable, give peace of mind and are a thrill to watch. They are easy to train if exposed young enough and their centuries- old genetic instinct to bond with livestock keeps them focused on deterring predators from harming the sheep.

Chapter 38
Landrace Breeds — A Result of Adaptation

To set the context for this chapter, let me define what constitutes a breed.

A breed is a group of domesticated animals sharing a common heritage that have been bred to pass on recognizable traits to their progeny. Some breeds show traits that were selected for by shepherds such as coat color, ear carriage and high milk production. Other breeds exhibit traits resulting from environmental adaption, forages, parasites and climate extremes, combine to shape local populations into locally adapted breeds. This process could be refereed to as propagation of the fittest.

Further breed definition has three categories.

1. Landrace breeds. Landraces are domestic populations shaped primarily by natural selection. They are well adapted to specific environments and have excellent reproductive and survival characteristics.

Landrace breeds generally have not been selected for uniform production and may well be variable in appearance color and size, but consistency is the rule in their ability to thrive and reproduce in specific demanding environments, basically without human intervention.

2. Standardized breeds. This class of breed developed likely as a result of sheep shows, judges and show ribbons from human efforts to organize landraces into breeds of uniform size, color and appearance.

Production traits and cosmetic points are selected toward a breed standard, hopefully resulting in uniformity of appearance and consistent, predictable production. Breed standards are a good thing provided they take adaptability into consideration and allow for desirable variants in different environments.

Many of the oldest standardized breeds of sheep such as the hill breeds of England began as landraces because environmental adaption used to matter.

3. Crossbreeds and composites. Sheep containing the blood of two or more breeds are called crossbred. Crossbreeding creates heterosis, hybrid vigor, the result hopefully being greater production at all levels.

All crossbreeding programs should be aimed toward producing terminal fat lambs. Crossbreeding two maternal breeds may give the resulting ewe lambs a boost of hybrid vigor, and they in turn can be bred to a paternal breed ram to optimize carcass traits and yield on the final lamb crop. These three-way cross lambs should all be harvested for the freezer.

Sometimes crossbreeding is used to combine the characteristics of two or more standardized breeds in an effort to form a new breed. The resulting group of sheep may be referred to as a composite breed. The Kathadin started as a composite breed that has been standardized.

Getting back to landrace breeds, the most important commonality in any given population is adaptability and reproduction. Individuals that are adaptable to their environment have a greater life expectancy. And individuals that possess traits that support reproduction will have more offsprings.

As natural selection advances, a landrace breed is born. Longevity, the duration of productive life will determine how many offspring an individual may contribute to the flock, thus longevity becomes the third hallmark of a well established landrace breed.

With ever rising production costs in order to maintain profitability, modern breeds of sheep may once again need to be selected and judged by the same criterion as their landrace ancestors, namely adaptability, reproduction and longevity. Incorporating those criterion into existing breed standards would add meaning to championships and show ribbons.

Chapter 39
Grassmaster Hair Sheep

Building our sheep business from the grass up in a land of extremes called Missouri, sheep adaptability has always been a front and center objective. After all, a shepherd only has three options.

1. Select adapted sheep.

2. Alter the environment.

3. Fail as a shepherd.

In 2012, one year after buying my first flock of pure-bred Katahdins, I became aware of a small population of landrace hair sheep indigenous to the neighboring county. The flock owners called their sheep Katahdin, but the gene pool had been closed for so long that no one really knew where the genetics originated. Only in 1999, an elderly shepherd who was dying of cancer asked his neighbor to take the sheep that he had tended the past 20 years and allow the flock to propagate.

Turned loose to shift for themselves, living and reproducing on a forage-only diet with very little interference, this landrace population of hair sheep had developed desirable characteristics that were not commonly found in the Kathadin breed, characteristics that I wished to incorporate into my purebred flock.

Remembering my good experience with the Gulf Coast native, a landrace wool breed, I decided to give these unknown hair sheep a try. For the sake of convenience and to make a fair comparison, they were co-mingled with my purebred Kathadins and later we added Dorpers, St. Croix and Barbados.

A four year trial was enough to convince me that the unknown landrace sheep were my best breed option. So we purchased all of the remaining ewes and began a rigorous breeding program, improving cosmetic points and uniformity of size while staying focused on adaptability, reproduction and longevity. During this time period we introduced some Katahdin blood to improve shedding ability and relieve inbreeding. Accomplishing those goals, we closed the gene pool in 2018.

Over forty years of selection by a local environment, coupled with my recent efforts to standardize appearance, resulted in a productive, albeit comely breed without a name. So in January 2019 as recognition for the sward that has sustained these genetics I dubbed this landrace gene pool Grassmaster Hair Sheep.

Secrets are hard to keep. Word started to get around about this old/new breed and the rest is history. Between 2019 and 2023, Grassmasters, mostly rams, have been sold to shepherds in many environments and have proven their adaptability in eleven states.

Basic Description of Grassmaster Hair Sheep
* Moderate frame. Ewes average 115 pounds in their work clothes.
* Individuals should give the impression of symmetry and balance.
* Heart girth to top line ratio — approximately one to one.
* Short, smooth summer coat, long dense winter coat.
* Shed their winter coat 100% annually without assistance.
* May be any color or combination of colors.
* Low maintenance, early maturing and parasite resilient.
* Alert and intelligent with a calm temperament.
* Rams are polled, may have small scurs, promient scurs are objectionable.

Grassmaster Standard for Performance
* Adaptability, no crutches needed. They thrive on quality forages.
* Reproduction, mature ewes should wean two lambs.
* Longevity, eight plus years of production.

Adaption to our Midwest winters has shaped the Grassmaster into a deep, thick, compact sheep with a dense winter coat. Back fat accumulated during the fall serves as insulation during cold weather, and indeed back fat is also an energy reserve that proves its value during the long months of winter. The same phenomena that makes these sheep adaptable to grazing all winter also creates a rotund, small frame fat lamb, much appreciated by Middle-Easterners.

I am making no effort to form a registry for this wonderful landrace breed. While we maintain strict production standards and continue selecting toward a breed standard to improve uniformity of appearance, I want nothing to do with registering individual sheep.

Registration ruins breeds. "Sheep Number 1735 can't be culled because she is the daughter of #1422 who is the best female in the breed." Sound familiar? Strict, indiscriminate culling is essential for the general health of any breed or gene pool, and commercial animals, not registered, are the easiest to cull.

Unfortunately stock shows designed to promote registered animals couldn't care less about adaptability or reproduction. Ribbons are awarded to unproven, sometimes sub-par animals based on their physical appearance. In turn, show champions are promoted as Best of the Breed. Please understand me, I am not criminalizing fun at the fair, but I am suggesting that show winning sheep likely won't excel in traits that are economically important to the commercial producer.

In a commercial flock every sheep must stand on its own four feet, no excuses, no pedigrees, just performance. The further we have gotten from show winning genetics, the more functional our sheep have become. That's why I like landrace breeds. Besides, performance records have far more value than pedigrees. Registration will never eliminate genetic misrepresentation. Buyers simply need to know the seller.

Since the ideal finish weight of a lamb is 2/3 of its dam's weight, filling the growing market niche for small fat lambs weighing 60-80 pounds suggests a mature ewe weight of 90-120 pounds.

Unfortunately, some of the most common breeds of hair sheep are selected to a standard that favors a 140-170 pound ewe. The result is a fat lamb that is too big for most Middle Eastern buyers and too small for traditional markets. To offset the maintenance costs of their bigger ewes, shepherds often turn to accelerated lambing. With this non-seasonal production model, labor, shelter and supplemental feeding requirements all increase. The conclusion is simple enough. Breed standards should be market oriented, assuming that orientation doesn't preclude environmental adaptation.

Being moderate framed, the Grassmaster allows us to fully capitalize on the growth potential of our lambs and sell them on a strong market in cool weather.

We still keep a flock of Katahdins, but with each passing generation the survivors appear more and more like their Grassmaster counterparts. After all, they are adapting to the same environment. As a side benefit, our Katahdin slaughter lambs are better adapted to the 21st century market than they were previously.

If consumers are willing to pay a premium for a familiar sized, small fat lamb, resembling the lambs in their country of origin, why not get market oriented and sell them what they want?

Choosing a local landrace breed has decreased our cost of production and increased the pounds of lamb weaned per acre. That's why we call them Grassmaster Hair Sheep.

Chapter 40
Market Orientation

Getting acquainted with market criteria, options and trends is time well spent. I refer to this process as market orientation. For shepherds, knowing what is up and what is down in the marketplace is just as important as knowing North from South and East from West. Markets are simply a function of supply and demand. Yes, they can be manipulated, but at the end of the day a shortage of desired product causes a rise in price and over supply will cause a dip in the market.

If you think lamb is sheep and sheep is lamb regardless of size and age you may want to hang out in your nearest sale barn for a few auctions and visit with the buyers.

Consumer preferences for size, sex and age of lambs or sheep varies not only by race but also the country and heritage consumers hail from. While some consumers are looking for lamb chops, leg of lamb and lamb burger, which is best derived from a large lamb to reduce processing costs, other consumers want a small whole lamb to celebrate various occasions. The whole lamb may include the head, testicles and organ meats.

It seems this diversity of consumer preferences developed in part from environmental pressures in our ancestors' countries of origin. A Middle Eastern chef once told me that every Middle Eastern population prefers a little different size and age lamb. The range preferences from small to large was astounding. The harvest age range between countries was equally as broad.

The same chef told me the variation in size and age at harvest stemmed from the type of sheep available in each region. Of course, regional differences in forage determined the type or breed of sheep that lived there. And this environment also affected how quickly a lamb could finish. A fat, 18-month-long yearling is preferable to a thin lamb, so if a country couldn't produce fat lambs in seven months they might opt to eat fat yearlings instead. This concept of breeding sheep adapted to one's region and food preferences formed by what is locally available is intriguing to me.

Here in the USA we have a wide range of climate and elevation. The diversity of sheep breeds and shepherds is equally dramatic. These conditions give shepherds an opportunity to produce the diversity that best matches our melting pot of cultures.

If cost control is of any interest to you, adapting production to your regional limitations should be first on every agenda. However, most environments allow enough flexibility for some adaptation to market trends. Using a meat breed as a terminal sire on smaller, easy keeping ewes is a reasonable way to produce big lambs if that is what your market wants.

Reading sheep books that were written in the 20th century gives the impression that every lamb needed to reach a finish weight at or above 120 pounds. That trend has changed considerably in the past 20 years. Today, a lot of lambs are being harvested between 55 and 80 pounds liveweight.

Another consideration for the 21st century is the demand for whole, unblemished lambs. Many cultures prefer intact ram lambs. Castration is considered a blemish. Hence the importance of market orientation.

While some cultural groups like sheep that are over a year old, most cultures claim to prefer "lamb." Age on live animals is usually determined by the teeth. If you are selling through the sale barn and your lambs have broken in their yearling teeth the price of your lambs will probably be docked heavily. Most lambs will get their yearling teeth between 11 and 13 months of age. Sometimes they won't break in until they are nearly two years old. There is nothing inferior about a 14 month "lamb." Rather than take a hit on yearlings that need to be repurposed because they failed to reproduce, we have been developing a market for older lambs among our friends and acquaintances.

I have never understood why the price should be docked on yearlings. Personally I prefer to eat a three- to four-year-old ewe that is healthy and fat. They have more flavor than lambs and many of them are still quite tender.

I suspect that a lot of the mutton sold on the open market is called lamb by the time it reaches the consumer. One year I gave in to my sentiments for woolies and bought an 80-pound Shetland ewe. She proved to be incapable of adapting to our ranch, got sick and lost a lot of weight. I took her to the sale barn. When she came in the ring she weighed 65 pounds. The owner of the establishment who worked in the ring and also bought about 25% of the stock that came through turned to me and asked, "Abram, she's not a lamb, right?"

I affirmed that was true. The owner of the sale barn proceeded to bid that Shetland ewe and bought her. And here is the punch line. As the little 65-pound ewe walked out the door I heard the new owner mutter under his breath, "She will be a lamb now."

What he didn't know was that underneath all her wool she was extremely thin. But 65-pound lambs were bringing $4.00/pound at the time and he bought that ewe for $2.00/pound. I wonder how she tasted.

Fat healthy cull ewes are worth far more as chops and burger than they will ever bring in the sale barn. Be honest about what you are selling, but don't discount your product. It's good meat. If you have some really sorry looking culls spare your customers a bad eating experience and send them to the sale barn or process them and market them as pet food. Doggie treats bring good money these days.

Seasonal price cycles are created by seasonal production and amplified by everyone wanting to sell their lamb crop at the same time. The result is a depressed market followed by a corresponding peak on the opposite side of the production year. For example, if most lambs are marketed May through

September then prices will generally be a lot higher December through March.

Within this seasonal price cycle there are little hiccups over Easter, Ramadan and other occasions. Some of these occasions are based on a lunar calendar resulting in the holiday coming 11 days sooner each year. This makes for a changing target that moves through all the seasons. The open market is often crowded leading to these holidays. As a result lamb prices are often higher after the holiday than they were before.

Stepping back from the seasonal price cycle we also have a multi-year price cycle. This cycle is also driven by supply and demand. As lamb prices rise, more producers are attracted to the business and those who are already in business retain more replacement ewes than normal. Often ewes that should be culled are retained for another year or two. With less sheep on the market, prices climb even higher. More sheep are retained and prices rise still higher.

Eventually the price cycle reaches its peak. Something bumps the trigger and the market comes crashing down. When prices start falling new shepherds who were attracted only by the sight of $$$$ sell all their sheep driving prices even lower. Widespread drought tripped the trigger on the most recent price cycle, which took about ten years to complete.

The important thing is to keep in mind that with the commodity sheep price cycle lamb prices are at the low part of the cycle, far longer than they are on the high side. Structure your business in a way that you still make a profit when prices are low. Spend your money wisely when prices are high and you will be able to stay in business for the long haul.

If you have the time and ability to market your lamb, fiber or sheep milk direct to the consumer you will have some reprieve from both the seasonal and multi-year price cycles.

Of course, to be successful at direct-to-consumer sales market orientation will be more important than ever.

Chapter 41
Our Global Neighbors and Controlling Costs

For the record, I am a proponent of local food commerce. I want to buy or barter the food I eat from someone I personally know. And I also aspire to sell or barter food we produce within our neighborhood.

Of course, there is nothing wrong with selling surplus product to other neighborhoods. When push comes to shove, the neighborhood that can produce a commodity at the least cost and with the greatest value will control the market. The following illustration demonstrates this concept.

Two men were hiking in the mountains. They walked around a big rock and came face to face with a grizzly bear. When one man took off running, the other man hollered after him, "What's the use? Don't you know bears can run faster than men?"

The first man replied, "I don't have to outrun the bear. I only need to outrun you!"

I am tired of hearing how we need more regulation on imported lamb and mutton, why there should be government subsidies for sheep producers, and suggestions that American shepherds can no longer produce a competitive product. Spare me the excuses. Sheep production is based on solar energy and water, no matter which side of the world you are on. Who wants a handout to stay in business? Not me! If my business is not competitive it needs to be restructured.

Old timers blame low wool prices for decimating the North American sheep industry. There is a lot of truth in that claim, but there is still a significant quantity of lamb consumed in this country. Wool requires a tremendous amount of energy to produce. If meat is what you really want to produce, breeding the wool off your flock would be an efficient move.

In mid-nineteen hundred there were more than 50 million sheep in the United States. As of this writing, numbers have fallen to approximately five million head. This awful drop in domestic production was probably precipitated by several factors. But the fact that imported lamb has been exceeding 200% of domestic production would suggest that our global neighbors are much more proficient or masterful at lamb production that USA producers at large.

Competition is good. Rather than whining, complaining and trying to hinder Australian and New Zealand imports of lamb, why not emulate their production models and beat them at their own game?

We are good at doing things efficiently, but are we doing the right things? Forage and labor are typically the greatest costs in the livestock business. Over the years we have tweaked our production model here at Still Waters to where labor is directed almost entirely towards reducing feed costs. This has been accomplished primarily through pasture management and controlling the breeding season. Not to offend anyone, but I would suggest that if you need to be on call 24 hours a day during lambing season, your business is not competitive. Maybe not even sustainable.

Greg Judy asked me to help teach his grazing school in May 19-20, 2023, which just happened to fall right in the middle of my normal lambing season. Ever confident in my ewes' maternal ability I accepted Greg's invitatiom. I gave the flocks each a two-day paddock and left the ranch for two days in the middle of lambing season. Of course, this action really excited a lot of other shepherds and even some who used to be shepherds. They were sure my "negligence" would end in dire consequences. Guess what, not a single ewe or lamb perished while I was gone!

Joel Salatin once told me about the time two of our

neighbors from Down Under came to visit his Polyface farm. While they were there Joel got to thinking and realized it was spring in the Southern Hemisphere. So he asked his guests, "What are you doing up here right now? Aren't you lambing yet?"

Their reply was, "Yes, and by the time we get back home, the ravens will have cleaned up the dead ones and we will be profitable!"

Obviously those guys were compulsive about controlling their labor costs.

As a shepherd, stories like the one mentioned above seem to be attracted to me by magnetic action. Here is another reportedly true account.

A New Zealander, who on a visit to a USA sheep ranch, was speechless when he learned that the owner needed an employee to help care for the few hundred head he owned.

Obviously irritated, the American turned on him and demanded, "Well, how many ewes do you lamb out?"

"Oh, just over 3000."

"How much help?"

"Just three dogs and me motor bike."

"Wow, how do you manage?"

"Well, I have a part time job in the winter to make ends meet."

Attitude and business structure can go a long, long way in controlling costs. Believe me, if you have $12.00 to spend there will be salesmen standing in line asking $13.00. The travesty is that so many purchase-able, fix-it-all products are addictive. Costly pest control leads to an ever greater need for more pest control. Just remember this, the best things in a sheep's life grow in the pasture and are FREE!

With forage being our greatest cost, my labor is almost exclusively aimed at pasture management to increase forage production and optimize forage utilization.

For a long time now shepherds have been trying with mixed success to increase output per ewe as a means of increasing profit. Accelerated lambing and triplet births have been front and center objectives. Those two objectives require a near perfect environment to be sustainable or else supplemental feed and shelter will be necessary.

When sheep prices hit bottom, all those inputs can put your enterprise in a risky position. The low end of the sheep price cycle often matches cost of production for high input producers.

One shepherd told me, "The only way to become wealthy in the sheep business is to spend less than you make." I have pondered his statement and concluded he is right. Unfortunately many people who hear that statement will yell, "Cheapskate!"

To run a profitable sheep enterprise we need to view our operation as a business that cannot be subsidized.

* If fertilizer is high, learn to manage your pastures in a way that eliminates the need for fertilizer.

* If grain is high, consider pasture finishing your lambs.

* If you are spending too much on labor, identify and eliminate time wasters.

* If forage costs are too high, reshape your production year to better utilize your cheapest forage.

* Be slow to kill something that wants to live and don't try to keep something alive that wants to die.

* Limiting production to your controlled resources may not sound exciting, but it certainly is the least risky and basically guarantees a profit every year. After all, who really wants to feel the bear's hot breath on their back?

I personally know ranchers here in Missouri, who in an effort to gain broader forage utilization, have added sheep to their cattle operation. Their only increase in labor is the need to control breeding/birthing seasons, and of course some time spent gathering and marketing their ram lambs and surplus ewes. The sheep are basically feral the rest of the year. Because the sheep are grazing forages underutilized by the cattle these outfits probably have a lower cost of production than most of our global neighbors.

I am not advocating feral sheep production. If sheep is your centerpiece, managing the flock will add dividends to your bottom line. Just be sure to leave the $8.00 an hour work for your sheep so that you are free to perform the $100 per hour jobs.

With a history of lamb and mutton imports exceeding 200% of domestic production annually, there is plenty of room for flock expansion, as long as you can make a profit in a depressed market. I am rooting for you.

Chapter 42
Capitalize on *Your* Overheads

Overhead costs limit the production of young, cash poor businesses. The title of this chapter reflects, "your" business needs to fully exploit your overhead costs.

Overheads for pasture based sheep operations can be divided into four basic categories, listed below:

1. Pasture. Whether we own pasture or lease it, pasture comes at a cost. If pasture availability is limiting your production and profit, before you shell out for more acres, optimize production on the acreage you are currently paying for.

Over time implementing adaptive grazing management can realistically improve production by 40%. That's going to be a lot less costly than buying or leasing a 40% increase in acreage.

Broaden your horizons. Land has many values besides pasture. Look for opportunities — recreation, hunting, agritourism. Here at Still Waters we capitalize on our land cost by selling educational farm tours to new and wannabe shepherds and well established shepherds looking for new ideas.

2. Livestock. Sheep, goats and cows all fall into one of two categories — either they are profitable or otherwise. However, they all have one thing in common. They compete for your pasture. Culling non-productive animals frees up capital and forage that can be invested into productive replacements. Culling a ewe that hasn't paid for herself may be painful, but keeping her around another year will only add to your misery.

Relevant to this topic is the choice of which grazing species to employ. In spite of the favorable economic impression that sheep and goats deliver, I have found cattle will generate more income per acre, if our available forages are best suited to the bovine species. In considering the sheep advantage we simply manage our pastures to propagate sheep friendly forages. If we lease a farm dominated by perennial grasses, cattle are employed for a year or two until the sward diversifies.

3. Labor. Pareto's Principle suggests that 80% of our efforts deliver 20% of our results. Not good. It is no secret that shepherds often spend 80% of the time coddling 20% of their flock. Not good.

Hiring help to birth, clean and assist suckling when the ewes could do that themselves is not good. Selecting sheep that are capable of living and reproducing withoutout your assistance is good. Spending your labor on pasture management lowers the cost of your pasture overheads. Now that is productive.

Let's face it, labor *is* an overhead cost, even if you are providing your own labor. Identify the jobs that produce the greatest results and spend 80% of your time in that capacity,

We are all limited to 168 hours in a week. Let's capitalize on our time.

4. Equipment. How much equipment do you really need to harvest and sell your forage on the hoof? When equipment is required, can you hire the job out? We all have limited equity and limited borrowing capacity. Tying up massive fortunes in equipment is a shame, unless that equipment is staying busy, generating a positive income.

For my size operation, I find it is far more economical to hire a neighbor to custom bale our hay than to own a haybaler myself. If I did own a baler we would offer our services to the neighborhood to capitalize on the overhead cost of the baler. Moral, the smaller the operation, the less equipment you can afford, unless you can hire that equipment out to an advantage.

Polywire and polyposts can go a long way in reducing the need for traditional equipment costs. Electric fence has a low depreciation rate and is easy to service.

Summary: overheads limit the function of young, cash poor businesses. Invest your money and time in productive assets and leverage those assets to optimize their return on investment.

Chapter 43
Maximizing *Your* Profit

Too many livestock enterprises rarely if ever show a true monetary profit. The business is subsidized by an off-farm income, inherited wealth and free labor from the family.

For the sake of producing the most, biggest and best off farm inputs are used indiscriminately, which maximizes the salesman's profit instead of your own.

Profit is the fundamental basis of every good business, and if you are going to be fully employed in your own business it had better be structured to generate a profit every year, even if the structure needs to be modified annually,

Profit is not determined by how much money you put in the bank. For instance, if severe drought forces you to sell 25% of your flock the resulting income should improve your bank balance but doesn't equal a business profit.

On the flip side, if you choose to retain all your ewe lambs as replacements your bank balance may not look very good, but your business has made a profit because its net worth has increased in the past year.

Every business has three options:

1. Cease operations (bankruptcy).

2. Subsidize operations.

3. Stay profitable.

There are three ways to increase profits:

1. Improve the gross margin per unit, by cutting costs or raising prices.

2. Increase the number of units produced.

3. Increase turnover.

Since value of product, minus cost of production equals profit or loss, enterprise models should be compared to determine which model or variation will yield the greatest profit from your forage resources. That exercise is known as Gross Margin Analysis, analyzing the profit potential of an enterprise.

Should you have a ewe/lamb, feeder/finisher or a multi-species enterprise? Whether to go value added or reduce inputs are decisions businesses must make independently. The Gross Profit Margin of each enterprise will tell you where to focus your efforts.

The salesman's wares might increase production/output and the gross value of your lamb crop, but that doesn't ensure a greater profit; however, it does insure a greater risk of losing money.

Profit doesn't just happen. Profit planning should happen every year.

While some years have been better than others, my sheep business has shown a profit every year for the past 12 years. The business has been built literally with its own profits and is now buying a land base for further production.

We have worked to maximize profits every year, which has resulted in annual adjustments in production and marketing. As a self-employed shepherd, I take profit very seriously.

Profit should come in several forms — emotional, ecological, physical and monetary. Healthy businesses will seek to optimize profit at each level.

Chapter 44
Keep Records Informative

The title of this chapter could be applied to any farming or ranching record system.

From forage inventory to livestock held for sale, data is useless, even harmful unless it supports quick, appropriate decisions that move your operation toward your goal. Too often we spend 80% of our time compiling data and 20% of our time deciding what the data means. I suggest that reversing those numbers would only bring positive results.

For this discussion I am going to stick pretty close to sheep record systems. If you can draw some lessons for other endeavors, great.

Flock records should be geared towards eliminating poor performers and freeloaders. Records build favor for sheep that perform well and stay productive for many years. Thus, records can serve two purposes. They aid culling decisions and can guide replacement breeding stock selection.

Developing a huge data base for every ewe is painfully unnecessary. In fact, most necessary records can be displayed on an ear tag.

If a ewe has no bag at weaning time she has failed to re-produce herself. The cause — failed pregnancy, poor maternal, or sick lambs — makes no difference. She still failed. That's all I need to know. No bag at weaning time is sufficient evidence to convict a ewe. To simplify the info system we just notch her ear tag and she is culled at the first opportunity.

If a sheep needs any special treatment — hoof care,

assistance at lambing, respiratory medication — not given the rest of the flock, her tag should be notched also.

Obviously a notched tag could mean anything but the destination is the same. The beauty in this information system is the elimination of paper records. On sale day we don't have to consult the chronicles of the flock to discover three ewes that need to be culled. Better yet, we don't have to read every ear tag to identify #105, #267, #489. When a notch is seen we know it's a cull without consulting the history book.

Best of all, when #105, #267, and #489 left the farm they took their dismal list of failures with them. The record book simply shows we sold three cull ewes who will no longer be competing for our best forage.

Using records to select breeding stock gets a little more complicated. If we are only keeping ewe lambs and the ewe flock is culled ruthlessly each year as described above, parentage verification is hardly worth the bother. However, if selecting ram lambs for reproduction, ideally maternal lineage would be traceable for several generations. Obviously this is going to require some records.

Reproductive success is my first interest. Did the candidate's mom lamb at one year of age? Has she produced two lambs every year since then? Adaptability of the bloodline is proven by keeping their ear tags free of notches.

Longevity of the bloodline expressed by ancestors' years in production is also useful informative data. Longevity proves adaptability and trumps reproductive extremes of younger ewes. What good is a 300% lambing rate if the ewe fails at five years of age? I would much rather own a ewe that produces healthy twins for eight years than stunted triplets for five years.

Records should be kept to the point of identifying freeloaders and crediting performance. Where #419 lambed and how much rain fell the night she birthed are largely immaterial. More helpful is how many lambs were born, what sexes, and a birth weight score — small, medium, or large. If you are good at weight estimation, birth weights can be recorded as approximate. I have done enough of it that we can usually get within half a pound.

Ear tags can display a lot of information reducing the need to consult the record book. If we tag a lamb at birth we use a blank tag and print its mom's number on the tag. This allows lambs to be paired directly back to their mom.

Any lambs retained as replacements are re-tagged when they are six to seven months old. I like small animal Z tags. They have the highest retention rate of any tag we have tried. The new addition to the flock is given her own identification number. This is written neatly on the front side of the tag. On the back side of the tag we include the year of birth (born in 2020 is abbreviated as 20, born in 2023 is abbreviated as 23). By writing the year of birth small and neat there is also room for her dam's number, if known, or the flock's number/initials she was born in. Other data such as single, twin, or triplet can also be recorded on the back of the tag. That code could register as 1, 2, 3.

It is best to apply ear tags in cool weather to allow the ear to fully heal before fly season. When tags must be applied in warm weather, applying pine tar at the site of the wound will repel flies. However, we have never had trouble with infections on newborn lambs tagged in hot weather.

Be creative and develop your own record system specific to your needs. Records are a great asset for decision making as long as they are informative and stay on point.

The late animal scientist, Jan Bonsma, once said, "It is an established fact that longevity (the duration of productive life) is a constitutional trait that varies in different breeds and families."

Let's use informative records to build longevity into our flocks of sheep.

Chapter 45
Time — Using It Productively

As mentioned earlier, well over a century ago Pareto, an economist from Italy concluded that 80% of our efforts produce 20% of our results. Ouch. Unfortunately the Pareto principle continues to manifest. Many shepherds spend 80% of their time coddling 20% of their flock.

Why not cull the 20% that demands 80% of your time and get more sheep? If hooves need trimming, trim them, then sell the offenders before the job needs to be repeated. The same principle can be applied to any isolated action needed to keep an individual sheep productive. In other words, if 20% of your flock needs to be dewormed, must lamb in jugs, or require supplemental feeding, sell them.

Manage for what you want. Without clear achievable goals we make very little progress. When we work on what we don't want (problems) the same issues will appear on the agenda time and again. The results — 80% of your time will be spent putting out fires leaving you very little time for improving your business.

Okay, so we are in a rut, got fires burning everywhere, not making any progress, but fire will only burn until it runs out of fuel, so let it burn and take time to refocus. Pull up a chair under a shade tree or make yourself comfortable on a bale of hay in the barn. Allow yourself time to think, ponder and meditate. This isn't lazy time. It's a highly productive exercise. Taking time for thought allows us to stay focused.

Develop good habits:

* Complete one job before starting the next.

* Avoid unnecessary interruptions.
* Keep desk and workbench clean.
* Have a place for everything and everything in its place.
* Take time to relax. Completely.
* Don't procrastinate. Be proactive.

Identify your time wasters, misplaced tools, junkmail, gossip, salesmen, TV, video games. Make a list and take action.

Records can be productive, but too often they are a waste of time. Spending 80% of your time compiling mountains of data and 20% of your time considering what the data means is not very productive. Reverse those numbers and records begin to look more like a meaningful information system.

As a shepherd you should be the CEO of your flock. Spend your time strategizing the flocks' actions and let your sheep employees do the work. With good management a flock of adapted sheep are the most reliable workforce you can imagine, never late, don't sue or go on strike. Yes, sheep are super good at producing the world's best food called lamb.

I frequently meet shepherds who wonder how I could possibly keep up with 400 ewes. On the flip side folks who come here for an educational farm tour and witness the rhythm of production wonder how I manage to stay busy. For the first question it's all a matter of attitude and structure of the business. For the second question, I do stay busy, but not with the sheep. Christina and I are raising a family, cultivating an orchard and acting as stewards for our little corner of creation. And sheep are not our only livestock. Yes, they are the centerpiece, but the goats, horses and cows all compete for our attention.

Plan your time. If your memory is a little fuzzy, write things down. I used to laugh at day planners, but not anymore. A dull pencil beats a sharp mind.

We have all been given 24 hours in a day. Let's use our time productively, taking time to think, ponder, and meditate may be a good place to start.

Chapter 46
Notes of Interest

1. Forbs for sheep and brush for goats are essential, just like grass is essential to the well being of cattle. Shepherds should tailor pasture management to utilize forbs without eradicating them.

2. New shepherds often enjoy wonderful sheep performance for two to three years until their flock has eradicated the forb component of the sward and become 100% grassfed by default. If cattle are managed in a way that eliminates grass from their diet, herd resilience will suffer. The same thing happens when sheep lose access to weeds.

3. Maintaining body condition is not proof that nutritional requirements are still being met. Sheep that gain and maintain weight easily may starve to death while still fat.

4. Nutrition may be effective as a remedy but is much more effective as a preventative with 80% of the immune system occurring in the gut. Forage and water quality have a major impact on how well the immune system responds to daily threats.

5. Natural is a subjective term when supplements are being discussed. For example we can blend two or more raw natural products and thus entice livestock to consume natural matter that is unnatural to their metabolism.

6. Weeds can be accumulator plants collecting and storing in their tissues up to several hundred times as much of a trace mineral as an equal quantity of soil contains. Each species also has a different mineral composition.

7. Minerals contained in plant life are generally 350%

more bio-available than minerals that are mined out of the earth and sold to you in a bag.

8. Set stocked animals will out perform animals on a managed grazing regime until forage diversity declines. Short term, set stocked sheep have maximum daily forage selection and consistent shelter and water options.

9. With forage diversity and quality being fundamental for sheep performance grazing should be managed in an effort to sustain or improve forage quality from day to day and year to year.

10. Co-grazing sheep and cows — should it be sheep first, cows first or both species together? That question frequently pops up at grazing conferences and pasture walks. As graziers we should be adaptable enough to use all three configurations depending on the varying nutritional needs of the flock and herd. For example, if the sheep are lactating and the cows are dry, send the sheep ahead of the cows.

11. Commodity lamb prices are generally at the low part of the price cycle for longer than they are on the high side. The low side of a price cycle often matches cost of production for high input producers.

12. Least cost producers can ride out price cycles by identifying and eliminating unnecessary costs from their production model. Just don't confuse least cost producers with cheapskates who try to starve a profit out of their flock.

13. To optimize tenderness, lamb and mutton carcasses need to be kept at ambient room temperature for a couple of hours after slaughter. The enzyme action that keeps muscles from shrinking and tightening and instead makes them relax only works for a couple of hours after the animal dies and is only viable at ambient room temperature. Consequently, rapid chilling results in tougher meat.

14. Dry weather optimizes animal performance so long as the forage stays green. Sustained cloudy, rainy weather is tough on livestock and especially sheep and goats.

15. Coccidiosis can occur any time of year in under-nourished or stressed sheep. Ram lambs followed by mature rams are more susceptible to coccidiosis than are ewes and ewe lambs.

16. Stress and malnutrition are the passwords that allow internal or external parasites to take out a sheep. Anything the shepherd can do to mitigate stress will improve resistance to parasites.

17. Hot muggy weather usually precedes Haemonchosis (parasites gone wild). Anemia is the most important symptom of Haemonchosis suggesting necessary action against the worms. Taking measures to keep the flock reasonably cool and comfortable will go a long way in preventing severe Haemonchosis. Allow access to shade, preferably shade trees.

18. Even though parasite resistance is thought to improve with age, young, therefore susceptible lambs are protected from parasitism by their mothers' milk, provided the ewes are strong and healthy.

19. Footrot and its precursor footscald is usually caused by poor circulation, stimulated by stress and amplified by unsanitary conditions such as mud.

20. Easy fleshing ewes who fail to raise lambs annually may accumulate fat in their udders. If a ewe has a large udder at lambing time but 10 days later her lambs appear to be starving she probably has an udder full of fat, thereby reducing her milking ability. There is no cure for fatty udders. Just cull the ewe.

21. The number of lambs weaned per ewe is much more important than how many lambs are born. I call this the lamb production percentage. Raising the production percentage is best accomplished by culling low performers rather than selecting for triplet births.

22. Weaning weight percent, the weight difference between a ewe and her weaned offspring, is the ultimate means of comparing annual performance between ewes of different sizes. The higher the weaning weight percent, the better a ewe has done. A ewe should be able to produce her own bodyweight in weaned lambs by the time they are four months old.

23. Longevity, the duration of productive life, is a

constitutional trait that varies in different breeds and families.

24. Selection is the mightiest tool in the hands of the shepherd. It is the primary means to bring about improvement and to mesh one's sheep with the environment they inhabit.

25. In general, a market animal is finished when it weighs 2/3 of the average mature female weight of dam and sire breeds. The range is 60% to 70% for livestock. Research at the University of California Davis found 64% for lambs.

26. Shrinkage. Thanks to loss of manure and urine, sheep will shrink when shipped. Shrinkage may vary from a common 3% to an extreme 10%. Most shrink occurs in the first 20 miles. Due to a greater gut capacity pasture fed lambs will shrink more than grain fed lambs.

27. Knowledge versus education. Through the process of observation, I learned years ago that sheep relish birdsfoot trefoil and horses dislike the same plant. That was knowledge. Recently, I learned that Birdsfoot trefoil is high in tannin, and that sheep aremore tolerant of tannin than horses. That was an education.

28. People who manage to stay in animal agriculture long term by nature are very patient, or at least learn to exercise in that capacity. Change and improvement take time. Rash decisions bring dire consequences. From conception to birth and birth to sale, patience is a virtue, a character trait of successful stockmen.

29. The difficulty of learning something new is not the new ideas but giving up old notions. If you want to make small changes, change the way you do things. If you need to make a radical difference, change the way you see things. Observation is fundamental to successful stockmanship. Let it guide your actions.

30. I'm not farming an inheritance and yet I have received a lot of agricultural value from the older generation — cooperative opportunities, sharing knowledge. Would I be a shepherd if 20 years ago my Dad had said, "No sheep on my land." Honestly, I don't know, but I do know the early exposure and opportunity to care for some sheep sped me on my way to becoming a successful shepherd. There have been many fatherly and grandfatherly figures in my life to whom I went for advice when we had a dilemma. We wouldn't be where we are without that generational wisdom. If you are a young person and you have an opportunity to pursue with an older partner be dependable.

31. Farming has always been a generational occupation, and always will be. Knowledge can be inherited to or from anyone, and is not taxable. Keeping young people engaged on the land is the best investment an elderly farmer can make. If the farmer wishes to take a few days off, reliable, knowledgeable help is already present to take up the slack.

Chapter 47
Shepherds Need Each Other

I'm not the only shepherd in step to the rhythms of nature, and I didn't learn everything I know the hard way. Without working examples to model, I might still be trying to hybridize confinement jug lambing with grass farming. I have learned from many shepherds and other stockmen who followed one or more of the principles I covered in this book. All those hardworking stewards of the land come from many backgrounds and different environments.

The point I want to drive home is this, you can learn something from almost everyone.

In that sense this book is a collective work of many, many people. Yes, I wrote it, but many others furnished the content in the form of working and failing examples, suggestions when my operation was struggling, and yes, even questions from wannabe shepherds. Questions provided the stimulus for this book. I hope you found your answers in these pages. Undoubtedly there will be more questions. I recommend developing a support team. Locate producers in your area and learn from them.

As mentioned earlier, when I was 15 my Dad allowed me to implement managed grazing with his set-stocked beef cows. The cows were not a farm priority. They were kept to eat grass and hay, thus utilizing the acres not involved in wholesale vegetable production.

I really didn't have a clue how to manage pasture or cows, but with an open mind and close observation it didn't take long to develop a protocol that seemed to suit everyone. If the cows were moved before they became excessively hungry there would be enough residual forage to protect the soil and serve as a solar collector to fuel regrowth.

With this scenario the cows were happy because there was plenty to eat. My Dad was happy because the cows were doing well. The grass was happy, which allowed soil life to proliferate. I was happy because of a sense of accomplishment and because seeing the improvement, my Dad allowed me to spend more time with the cows. Remember? Growing tomatoes wasn't my thing. I had long since memorized the number of small, medium, or large tomatoes needed to fill a 20-pound case. Yes, manufacturing tomatoes was just too narrow a discipline for my active mind.

After a couple of years of trial and error with my Dad's cows and pasture, a neighbor noticed my efforts and invited me to accompany him to a Green Hills Farm Project (GHFP) pasture walk hosted by Dennis MacDonald. Since that day Dennis has become a close friend and mentor. I probably wouldn't have written this book if I had never met Dennis. Indeed, he likes to remind me that *he* taught me everything I know. Thank you, Dennis. That was my introduction to GHFP.

GHFP members voluntarily take turns hosting a pasture walk once a month April through October. A round table discussion is held in January and a seminar in February. GHFP charges a membership fee to generate money for the winter seminar. At the pasture walks a potluck dinner is served (everyone contributes). After all, food is a part of community. The pasture walk may be conducted before or after the meal. Out in the pasture the discussion can go about anywhere.

At that first pasture walk I was invited to attend, I learned about leaf counts, manure quality, gut fill, portable freeze-proof water, forage mass accounting and more. I saw new fencing techniques, met a lot of competent graziers and listened in on conversations and debates ranging from basics to advanced. Obviously the larger and more diverse the grazing group, the more exciting it gets!

GHFP and Dennis MacDonald gave me a lot to think

about that day. Years later I am still learning, still refining ideas and yes, now I'm passing along some of what I learned.

Is there a basic level needed to succeed? Probably so, but assuming that you know it all will set you up for disaster. Attending one pasture walk or reading two books may set you on the right path, but you won't know it all. No one knows it all and as soon as we think we do nature will take us down a few notches.

That puts me in mind of the time I was asked to speak at a two-day grazing conference. At the end of the second day when all the speakers were called up front for a wrap up session of Q & A a young fellow who was considering grass farming as a new career posted a question. "Now that we have all this info, seems like we could go out and do just about anything. How do you stay humble?"

The answer he received was "The land will keep you humble." That solemn remark was echoed by all the speakers present, whose main occupation was farming.

Learning from others' mistakes is cheaper than your own. The GHFP serves as an example of isolated graziers banding together to pool their knowledge and share ideas. Coordination and communication are the sole costs required for the group to meet on a farm each month. Participation is really the limiting factor that hampers the benefit and success of GHFP and similar organizations.

Another relationship used to perpetuate eco-friendly livestock production is consulting. Of all the variations, on farm consulting by actual practitioners seems to be the most effective. Obviously a cash exchange is what keeps this model of learning mutually beneficial.

Apprentice/mentorship programs, topic relevant conferences and seminars, or well moderated discussion groups all are viable ways to learn. I like to surround myself with folks who are open, honest, welcoming, diverse and who operate in an "iron sharpens iron" mentality.

Personally I have learned a great deal from books. They are perhaps the surest means to perpetuate information from generation to generation. It has been said, "If you want a new idea, read an old book." So true. The Bible, God's book, gives us a good picture of how important animals were to our ancestors and would bear out that nothing has changed in that regard. The Old Testament gives us a glimpse of ancient stockman knowledge and ability seemingly far exceeding anything we have today, in spite of all our scientific advancement.

Reader driven magazines are a great way to keep up with current thought and innovation, rediscovery. I've gotten numerous ideas from *The Stockman Grass Farmer.* Applying them to our operation, some were highly successful, others we discarded. My point is, if we had waited for those ideas to appear in a textbook our operation would be years behind the rediscovery of the nuances of holism.

Why do people need each other? It has been said that human relationships ultimately determine our level of success. Let's quantify that statement. "Connections with other people determine our level of success." Some of my closest connections are immediate family, neighbors, customers, product suppliers, friends and mentors.

How many people do you need on your team? Enough to be successful!

If you want to go fast, go alone, If you want to go far, go together. People need each other.

Chapter 48
Staying Focused — Tying it All Together

Circumstances are always changing — forage species, markets, weather patterns, economy, land availability. The only consistent thing is change!

Taking time for observation, staying focused and adaptable is the key to survival in any business. Benjamin Franklin once said, "The eye of the master will do more work than both his hands." I reckon Benjamin was trying to tell us that without observation the master's hands will be working on the wrong projects.

I have attempted to break this book into 48 chapters to facilitate reading, yet I find the subjects remain interwoven. The effort required to divide subject from subject has made me realize just how connected all life really is. Yet shepherding is quite simple, provided we stay focused on the six principles listed below.

1. Forage suitable to grazing species.
2. Stocking rate in balance with carrying capacity.
3. Propagate adapted genetics.
4. Time breeding seasons to optimize reproduction and survival.
5. Adequate shelter/protection.
6. Gut friendly conditions.

I have never seen these six principles connected as a template for healthy, happy livestock. In our modern world we have convenient access to crutches that attempt to replace each of these principles. But crutches come at a cost. That cost is unhappy livestock, increased workload, high cost production, compromised food and a degraded environment. In buying these needless crutches we allow the farm income to be siphoned off by helpful salesmen and huge corporations.

While some sheep are certainly more resilient than others, and mine are no exception in that regard, no sheep is adaptable to drinking dirty water, living on injectable vitamins and sustaining themselves solely with inorganic minerals.

I've tried it and yes, that's right, got off mission and paid dearly for the experience. Lost my focus so to speak by overlooking principles 1, 2 and 6, thinking that performance could be sustained with purchased nutrition contained in bags and bottles, believing water quality didn't matter provided the sheep were forced to consume a natural detoxifying substance that shall go nameless (they say it works).

When finally as the prodigal son, I came to my senses, I realized the flock had been trying to catch my attention from the start. Yes, they wanted my focus. Their noise, performance and hair coats all screamed for change — plenty of sheep-friendly forage and naturally clean water.

If we don't stay focused, change will come by and by. First the sheep will fail to thrive. Next reproduction will suffer. Finally there will be a dead flock. That realization stimulated me to write the poem, "A Sheep's Petition."

As the poem says, sheep can be quite productive and low maintenance when we do our part as the shepherd. The best things in a sheep's life are found in the pasture. That is why sheep contained in a dry lot, force fed a scientifically balanced ration, are happy when they escape. And sheep that are contained on a manicured lawn are happy when they find their way through the fence into the neighbor's weedy pasture.

Stay focused, allow observation to guide your actions and enjoy your role. Career isn't destiny, it's process. Owning sheep doesn't make you a shepherd anymore than owning a farm makes you a farmer.

A shepherd/shepherdess is one who cares for and manages his or her flock. Using very reasonable means within context to promote the welfare of the sheep and the landscape they inhabit. All this is in an effort to sustain or, better yet, improve life. I find the shepherd's career second to none.

A Sheep's Petition

Oh give me a zone
Where the water is clean,
Shade trees abound
And the forage is green!

Give me a shepherd
Who will mitigate stress,
Free-choice sea salt,
Leave the winter for rest.

Reliable guard dogs
To keep coyotes at bay,
Many weeds to eat,
Thus healthy I'll stay!

Somethings way wrong,
If my hooves need trimmed,
Difficult births,
Or new moms need penned.

My hide hair and hooves
Have a story to tell,
Hair on the poll
May serve as a knell.

Well fed and protected,
Two lambs I'll produce,
Years of productive
Low maintenance use.

While forages grow
And rivers still flow
From East to West
Sheep like pasture the best!

Shepherd Terminology

When I was just a little boy I remember a neighbor stopping by to let us know that a grain elevator owner was penned in the jug for a grain trading scandal. I asked my Dad what that meant and he replied that the elevator owner was put in jail.

Later on I discovered that the little pens sheep are expected to lamb in are called lambing jugs. So there, if your sheep lamb in jugs they are serving jail time. Ha!

Gradually, I began to realize that the sheep world has its own lingo. Some of the terms are abbreviations, other words are descriptive and to the point.

Below is a glossary of shepherd terminology. If you are new to sheep take the tine to get familiar with sheep lingo.

Afterbirth. The placenta and membranes that a ewe releases shortly after lambing.

Anemia. A condition marked by pallor and weakness. In sheep, anemia is most often caused by loss of blood due to blood feeding parasites. Anemia causes normally pink skin to appear pale.

Bag. The ewe's udder, mammary glands.

Bloat. A rumen disorder caused by abnormal accumulation of gas. Bloat is often fatal.

Body Condition Score. Also BCS. A system of noting the condition of sheep. Scores range from 1 (thin) to 5 (fat).

Broken Mouth. An old sheep that has lost one or more of its mature teeth. Tooth loss can begin as young as five years.

Buck. An intact male sheep, also called a ram.

CAFO. Concentrated animal feeding operation. A feedlot.

Carrying Capacity. The number of animals a piece of land can sustain from year to year.

Dock or Tail Docking. The practice of cutting off the tail.

Drift Lambing. A pasture lambing model that allows new pairs to stay where they lambed while the rest of the flock moves forward to fresh pasture.

Dry Ewe. A non-lactating ewe.

Ewe. A female sheep.

Flushing. Increasing the flocks' plane of nutrition to improve fertility.

Fly-stike. Living sheep infected with fly eggs or maggots.

Footrot or Hoofrot. A condition of the feet probably caused by poor circulation.

Footscald. A precursor to footrot though perhaps not as detrimental.

Forbs. The general term applied to broadleaf plants that are not woody, not a grass, and not a sedge. Forbs are often referred to as weeds.

Freshening. The stage of giving birth.

Gestation. The period when the young are carried in the uterus, pregnancy. Gestation length for sheep is usually 147 to 153 days.

Grazer. An animal that eats grass.

Grazier. A human who manages grazing animals.

Gummer. An old sheep missing all of its teeth.

Jug. A small pen usually 4 x 5 or 5 x 5 used to pen a ewe and her newborn lambs for several days after lambing.

Lactation. Yielding of milk by the mammary gland. The act of giving suck.

Lamb. An immature sheep, generally younger than one year of age.

Lambing %. A 190% would mean that the ewes birthed an average of 1.9 lambs.

Lambing Bed. The area a ewe chooses to lamb. Once her water breaks, she probably won't move from that spot for 12 hours for every lamb born.

Lamb Thief. A ewe that kidnaps lambs before her own are born.

Lanolin. The natural grease that a sheep produces to coat their wool or hair. Lanolin serves as a water repellant helping to keep a sheep's skin dry.

LGD. Stands for livestock guardian dog. LGDs are special dogs bred and selected for their instinct to bond with and protect livestock.

Mastitis. Infection of the udder or bag.

Polled. Naturally without horns.

Poll. The crown of the head.

Ram. An intact male sheep.

Replacement. A young sheep selected to join the breeding flock.

Retained Placenta. Placenta that does not pass, is not released by the ewe in a timely manner.

Ruminant. Animals such as cows, sheep and goats with a four-compartment system.

Scours. Unusually runny manure. Diarrhea.

Shearing. Clipping the wool off a sheep.

Shrink/shrinkage. The weight a sheep loses during adverse conditions such as during transport. Shrink is caused by a loss of manure and urine.

Sward. The grassy surface of the land.

Sire. A sheep's male parent.

Sound. Free from flaw or defect.

Teaser. A male sheep incapable of breeding ewes typically used to find sheep that are in heat.

Trigger Species. Plant species in the pasture sward that when grazed triggers a paddock shift.

Udder. A sheep's milk bag.

Weaning Weight %. The difference in weight between a ewe and her offspring at weaning time. If a ewe weighs 120 pounds and weans two lambs averaging 60 pounds each, she has weaned 100% of her body weight.

Wether. A castrated male sheep of any age.

Yearling. A ewe, ram or wether between one and two years old.

Author's Bio

Abram Bowerman is a first generation shepherd. He bought his first sheep at the age of nine with money he earned. Starting from scratch, he built his business from the grass up primarily on land he didn't own, utilizing alternative forage opportunities and grazing sheep for a share of the crop. He became fully employed as a stockman at the age of 25. Since that time, efforts have been directed toward buying land, but collaborating with cattle farmers provided a land base and opened the door to stockmanship.

Together with his wife, Christina, and their four children, they manage sheep, goats, cows, horses and poultry on Still Waters Farm in the Green Hills of Missouri near Spickard, utilizing 400 acres on leased and owned pastureland.

The Practical Shepherd celebrates over 20 years since buying his first sheep. Through trial, error and successes, the author has come to believe the best things in a sheep's life can be found in the pasture. With sheep as the centerpiece of the operation, Abram promotes low labor, practical solutions for challenges facing shepherds. His goal has been to raise happy, healthy livestock.

Living off-grid, Abram and Christina have no personal phones, computers, email or internet. Horses are employed for farming and local transportation. For long distance travel they hire a chauffeur or use public transportation. The best way to contact Abram is by snail mail, which takes three to five days longer than email. Write to him at 543 NE 90th Street, Spickard, MO 64679.

Acknowledgments

I would like to thank all the shepherds, practical and otherwise, who have influenced my thinking and provided examples of what works and what doesn't.

A special thanks to the *Stockman Grass Farmer* magazine for all the practical information. It has been a big boost.

To my Mother, who home schooled me. Thank you for your persistence in teaching me to read and write when I would have preferred to daydream or play in the pasture. Without that education this book wouldn't exist.

To my Daddy, who provded an opportunity to manage a sheep enterprise when I was a young boy. Thank you.

To Christina, my wife and business partner of nine years. Thank you for your steadfast encouragement through trials, experiments, errors and successes.

Last of all, thank you to all the graziers for your part in building the grass farming community. Graziers need each other.

Index

Questions
about grazing ???????
Answers *Free!*

While supplies last, you can receive a Sample issue designed to answer many of your questions. Topics include:

* Joel Salatin's Meadow Talk
* Birthing in Sync with Forage Supply
* Inflation Proof Your Farm
* Setting up Fence and Water Points
* Forage Chains
* Three Rules of Adaptive Grazing
* Pregnancy Testing
* The Economist Farmer
* Working with Border Collies
* And more

Green Park Press books and the ***Stockman Grass Farmer*** magazine are devoted solely to the art and science of turning pastureland into profits through the use of animals as nature's harvesters. To order a free sample copy of the magazine or to purchase other **Green Park Press** titles:

P.O. Box 2300, Ridgeland, MS 39158-2300
1-800-748-9808/601-853-1861
Visit our website at: www.stockmangrassfarmer.com
E-mail: sgfsample@aol.com

More from Green Park Press

COMEBACK FARMS, Rejuvenating soils, pastures and profits with livestock grazing management by Greg Judy. Grazing on leased land with cattle, sheep, goats, and pigs. High Density Grazing, fencing systems, grass-genetic cattle, parasite-resistant sheep. 280 pages. **$29.00***

CREATING A FAMILY BUSINESS, From contemplation to maturity by Allan Nation. Written with small, family businesses in mind. How to work with your spouse and children, emplyees and partners. Pre-start-up, pricing, production, finance and marketing. For anyone who wants to own their own business. 280 pages. **$35.00***

DROUGHT, Managing for it, surviving, & profiting from it by Anibal Pordomingo. Forages and strategies to minimize and survive and profit from drought. 74 pages. **$18.00***

GRASSFED TO FINISH, A production guide to Gourmet Grass-finished Beef by Allan Nation. How to create a year-around forage chain of grasses and legumes to create tender, flavorful grassfed products all year long virtually everywhere in North America. 304 pages. **$33.00***

KEEPING IT GREEN, A handbook for creating and managing irrigated pasture by Jim Gerrish. Covers all types of irrigation with pros and cons of each. Includes economics to determine best value and fit for your livestock class. 96 pages **$20.00***

KICK THE HAY HABIT, A practical guide to year-around grazing by Jim Gerrish. How to eliminate the most costly expense in operations anywhere in North America. Gerrish shares his experience gained in Missouri and Idaho. 224 pages. **$27.00*** Audio version: 6 CDs with charts & figures. **$43.00**

KNOWLEDGE RICH RANCHING by Allan Nation. Reveals secrets of high profit grass farms and ranches. Explains family and business structures. Nation shares knowledge he gathered over 30 years from successful financial ranchers worldwide. Anyone who has profit as their goal will benefit from this book. 336 pages. **$32.00***

LAND, LIVESTOCK & LIFE, A grazier's guide to finance by Allan Nation. How to separate land from a livestock business, make money on leased land by custom grazing, and how to create a quality lifestyle on the farm. Covers land-based financial issues to protect yourself from falling real estate prices without selling your farm or ranch. 224 pages. **$25.00***

MANAGEMENT-INTENSIVE GRAZING, The Grassroots of Grass Farming by Jim Gerrish. Takes graziers step by step through the MiG system. Using vivid images and detailed explanations, Gerrish begins with the soil, and advances through the management of pastures and animals. MiG basics: the power of stock density, extending the grazing season with annual forages. Chapter summaries include tips for putting each lesson to work. 320 pages. **$31.00***

MARKETING GRASSFED PRODUCTS PROFITABLY by Carolyn Nation. From farmers' markets to farm stores and beyond. Pricing, marketing plans, buyers' clubs, tips for working with men and women customers, and how to capitalize on public relations without investing in advertising. 368 pages. **$28.50***

NO RISK RANCHING, Custom Grazing on Leased Land by Greg Judy. Based on first-hand experience, Judy explains how by custom grazing on leased land he was able to pay for his entire farm and home loan within three years. How to find idle land to lease, calculate the cost of the lease, develop good water and fencing on leased land. Includes contract examples. 240 pages. **$28.00***

PADDOCK SHIFT, Revised Edition Drawn from Al's Obs, Changing Views on Grassland Farming by Allan Nation. A collection of timeless Al's Obs. 176 pages. **$20.00***

PASTURE PROFITS WITH STOCKER CATTLE by Allan Nation. Profiles Gordon Hazard, who stocked a 3000-acre grass farm solely from retained stocker profits and no bank leverage. Economic theories are backed by real life budgets, including one showing how to double your money in a year by investing in stocker cattle. 224 pages **$24.95*** or Abridged audio 6 CDs. **$40.00**

QUALTIY PASTURE, How to create it, manage it, and profit from it by Allan Nation, Revised and updated by Jim Gerrish. Offers low-cost tactics to create high energy pasture to reduce or eliminate expensive inputs or purchased feeds. How to match pasture quality to livestock class, stocking rates for seasonal dairying, beef production, and multi-species grazing. Includes how to create a drought management plan. 300 pages. **$30.00***

THE CALENDAR OF THE YEAR-ROUND GRAZIER by Steven Kenyon. Although Kenyon only has a 4 month grazing season, he explains how to graze year-round in any region. Includes economic and financial tips for profitability. 80 pages. **$18.00***

THE MOVING FEAST, A cultural history of the heritage foods of Southeast Mississippi by Allan Nation. How using the organic techniques from 150 years ago for food crops, trees and livestock can be produced in the South today. 140 pages. **$20.00***

THE PRACTICAL SHEPHERD, Trials, errors and successes while maintaining profitability by Abram Bowerman. Pitfalls and challenges to become a profitable, successful shepherd. What to look for when buying sheep, a calendar for breeding, lambing and weaning, managing pastures, when to make a paddock shift and economics. 216 pages **$25.00***

THE USE OF STORED FORAGES WITH STOCKER AND GRASS-FINISHED CATTLE. by Anibal Pordomingo. Helps determine when and how to feed stored forages. 58 pages. **$18.00***

* All books softcover. Prices do not include shipping & handling.

Name _____

Address _____

City _____

State/Province_____Zip/Postal Code _____

Phone _____

Quantity	Title	Price Each	Sub Total
____	**Comeback Farms** (weight 1 lb)	**$29.00**	_____
____	**Creating a Family Business** (weight 1 lb)	**$35.00**	_____
____	**Drought (weight 1/2 lb)**	**$18.00**	_____
____	**Grassfed to Finish** (weight 1 lb)	**$33.00**	_____
____	**Keeping It Green** (weight 1 lb)	**$20.00**	_____
____	**Kick the Hay Habit** (weight 1 lb)	**$27.00**	_____
____	**Kick the Hay Habit Audio - 6 CDs**	**$43.00**	_____
____	**Knowledge Rich Ranching** (wt 1½ lb)	**$32.00**	_____
____	**Land, Livestock & Life** (weight 1 lb)	**$25.00**	_____
____	**Management-intensive Grazing** (wt 1 lb)	**$31.00**	_____
____	**Marketing Grassfed Products Profitably** (1½)	**$28.50**	_____
____	**No Risk Ranching** (weight 1 lb)	**$28.00**	_____
____	**Paddock Shift** (weight 1 lb)	**$20.00**	_____
____	**Pa$ture Profit$ with Stocker Cattle** (1 lb)	**$24.95**	_____
____	**Pa$ture Profit abridged Audio -- 6 CDs**	**$40.00**	_____
____	**Quality Pasture** (weight 1 lb)	**$30.00**	_____
____	**The Calendar of Year-round Grazing**	**$18.00**	_____
____	**The Moving Feast** (weight 1 lb)	**$20.00**	_____
____	**The Practical Shepherd** (weight 1 lb)	**$25.00**	_____
____	**The Use of Stored Forages** (weight 1/2 lb)	**$18.00**	_____
____	Free Sample Copy ***Stockman Grass Farmer*** magazine		_____

Sub Total _____

Shipping	Amount
1/2 lb	$3.40
1-2 lbs	$6.00
2-3 lbs	$7.00
3-4 lbs	$8.00
4-5 lbs	$9.60
5-6 lbs	$11.50
6-8 lbs	$15.25
8-10 lbs	$18.50

Mississippi residents add 7% Sales Tax _____

Postage & handling _____

TOTAL _____

Over 10 lbs or outside USA call for postage.

Please make checks payable to

Stockman Grass Farmer
PO Box 2300
Ridgeland, MS 39158-2300

1-800-748-9808
or 601-853-1861
FAX 601-853-8087